W0017360

LIQUID-PHASE SINTERING

SPEKANIE V PRISUTSTVII ZHIDKOI METALLICHESKOI FAZY

СПЕКАНИЕ В ПРИСУТСТВИИ ЖИДКОЙ МЕТАЛЛИЧЕСКОЙ ФАЗЫ

LIQUID-PHASE SINTERING

V. N. Eremenko, Yu. V. Naidich, and I. A. Lavrinenko
Institute of Materials Science
Academy of Sciences of the Ukrainian SSR
Kiev, Ukrainian SSR

Translated from Russian

 Springer Science+Business Media, LLC 1970

Professor *Valentin Nikiforovich Eremenko*, a corresponding member of the Academy of Sciences of the Ukrainian SSR, is a specialist in the physics of alloys and in particular their surface properties. A graduate of Kharkov University, he attained his doctorate in 1960.

Yurii Vladimirovich Naidich is a graduate of Kiev Polytechnic Institute and attained his doctorate in 1969. His special fields are the physical chemistry of capillary phenomena and the wettability of high-temperature alloys.

Irina Aleksandrovna Lavrinenko was graduated from Kiev State University and achieved the degree of candidate in chemistry in 1966. Her field of study is powder metallurgy and surface phenomena, with particular emphasis on the sintering process of liquid metal phases and the effect of capillary phenomena on this process.

The original Russian text, published for the Institute of Materials Science of the Academy of Sciences of the Ukrainian SSR by Naukova Dumka in Kiev in 1968, has been corrected by the authors for the present edition. The translation is published under an agreement with Mezhdunarodnaya Kniga, the Soviet book export agency.

Валентин Никифорович Еременко, Юрий Владимирович Найдич,
Ирина Александровна Лавриненко
СПЕКАНИЕ В ПРИСУТСТВИИ ЖИДКОЙ МЕТАЛЛИЧЕСКОЙ ФАЗЫ

Library of Congress Catalog Card Number 78-107537
SBN 306-10839-9

ISBN 978-1-4757-5667-8   ISBN 978-1-4757-5665-4 (eBook)
DOI 10.1007/978-1-4757-5665-4

© 1970 Springer Science+Business Media New York
Ursprünglich erschienen bei Consultants Bureau, New York 1970.

All rights reserved

No part of this publication may be reproduced in any
form without written permission from the publisher

PREFACE

Industrial advances frequently depend on the development of new, special-purpose materials possessing specific magnetic, electrical, optical, strength, friction, antifriction, and other properties. Metal alloys produced by the conventional technique of metallurgical reduction often do not meet these new requirements. Powder metallurgy, therefore, is of considerable importance in solving many problems of present-day materials science. Its production techniques—solid-phase and liquid-phase sintering, impregnation, hot pressing — make it possible to obtain materials from metallic components which are immiscible in the liquid state and also materials in which metals are combined with nonmetallic components such as refractory compounds — oxides, carbides, nitrides, borides, silicides, sulfides, etc. The properties of sintered parts depend essentially on the processes occurring during their formation.

One of the most promising methods of producing sintered materials of high density with the best combination of various properties is liquid-phase sintering. In recent years, many publications have appeared concerning processes of sintering specific combinations of components, the theoretical basis of liquid-phase sintering, and the laws governing this process. The present work examines liquid-phase sintering processes and the action of capillary forces in models of dispersed solid—liquid systems, and also gives data from theoretical and experimental studies of liquid-phase sintering in various metal and metal—ceramic systems. Some theoretical generalizations on the principles of sintering processes are presented, and the driving forces of sintering and the effect of different conditions on liquid-phase sintering processes are considered.

CONTENTS

CHAPTER 1

GENERAL PRINCIPLES OF SINTERING IN
THE PRESENCE OF A LIQUID METALLIC PHASE

1. Role of a Liquid Phase in the Densification Process

Sintering is a complex process consisting of a series of closely related physicochemical phenomena. From the thermodynamic viewpoint, the reactions occurring during solid-phase sintering are similar to ordinary chemical reactions in condensed systems. Sintering of a pure single-component metal powder can be considered a simple reaction in the solid state. The propagation of the sintering process can be followed from changes in the physical properties of the part undergoing sintering and its dimensions. An important sintering criterion for multicomponent systems is the completeness of the interactions of their components, with formation of alloys.

The characteristic features of highly dispersed systems are the excess free energy resulting from the highly developed surface of powders and porous substances and the crystal structure imperfection resulting from the nonequilibrium condition of powders, this in turn being due to the particular conditions of powder production. As the result of shrinkage, stress relief, elimination of distortions, and reduction of the surface during sintering, the excess free energy decreases and the system becomes more stable. Thus, sintering should be regarded as a thermodynamic process in which the system tends to attain a state with minimal free energy.

One of the principal goals of studying the sintering process is to clarify the nature of the forces causing the drawing together of the particles, strengthening of the interparticle contacts, and, consequently, densification of a pressed or freely poured powder. Depending on the type of interaction between the components and process conditions, the sintering of mixed conglomerates of powders may occur in the solid phase or in the presence of a liquid phase. Studies have shown that the controlling processes in solid-phase sintering are diffusion, viscous and plastic flow, and recrystallization.

The diffusion rate of the components increases considerably in liquid-phase sintering, facilitating the displacement of solid particles with respect to each other, which results in rapid filling of pores and capillaries. In liquid-phase sintering the theoretical density of the material being sintered may be attained in a short time, since wetting of the solid particles by the liquid between them results in curved liquid-meniscus surfaces on which capillary forces act, tending to draw the particles together. It can be considered that the material being sintered is under uniform hydrostatic pressure. With good wetting the liquid infiltrates to the contacting sections of the solid particles, which sharply reduces friction and wedging. The particles are drawn together, the voids and "bridges" between them disappear, and thus the volume of the sintered material decreases. This shrinking process occurs fairly rapidly, its duration — a few minutes or tens of minutes — depending on the conditions.

In addition, the liquid phase, in which the solid phase is dissolved, intensifies the transport of the latter. In this case the mobility of defects at the particle boundaries may change; new imperfections in the crystal can be generated or the stressed state of the solid-particle lattice can be intensified under the influence of the wetting liquid (disjoining pressure). For this reason the sintering rate also increases sharply.

2. Processes Occurring during Liquid-Phase Sintering

The theory of liquid-phase sintering processes is usually based on the assumption that the liquid phase wets the surface of the solid particles and the solid substance is dissolved in the liquid.

The principles of liquid-phase sintering were first formulated by Price, Smithells, and Williams [105] in a study of sintered alloys of the W–Ni–Cu system. This system is characterized by the fact that the refractory component — tungsten — is slightly soluble in the copper – nickel melt produced during sintering.

These investigators noted [105] that the sintering mechanism in such a system, which was later called the heavy-alloy sintering mechanism, consists in the formation of a liquid phase that is present throughout the sintering process and an increase in the size of the solid-phase particles. The latter was explained by the tendency of the substance to be redistributed among particles with differing radii of curvature through the medium in which the substance is dissolved. The solubility of a substance increases with decreasing radius of curvature of its particles. For spherical surfaces

$$\ln \frac{C}{C_0} = \frac{2\sigma_{S\text{-}L} V_0}{rRT},$$
(1)

where C/C_0 is the ratio of the solubility of fine particles with a radius r and large particles with a radius $r_0 \to \infty$ (solubility of a substance with plane interfaces), $\sigma_{S\text{-}L}$ is the specific free surface energy at the solid–liquid interface, V_0 is the molar volume, R is the gas constant, and T is absolute temperature.

Thus, the fine particles are gradually reduced in size during sintering and dissolve in the liquid phase, while at the same time, due to the lower solubility of the large particles, an excess of the substance in solution is reprecipitated on the large particles, thereby still further increasing their size. The growth of the large particles continues until all the fine particles disappear and the structure of the alloy becomes relatively uniform.

The shrinkage of these systems during the initial stage of sintering is explained by the redistribution and concentration of particles toward the center of the compact under the influence of the surface tension of the liquid [105].

The theory of liquid-phase sintering was later developed by Bal'shin [4] and Lenel [94], who proposed two variants of the process:

1. The solubility of the liquid phase in the solid phase is comparatively small (the system is heterogeneous during the entire sintering period).

2. The solubility of the liquid phase in the solid phase at sintering temperature is so high that during the sintering process the liquid phase gradually decreases in quantity and after a certain time disappears completely. The system becomes homogeneous.

According to Bal'shin [4] and Lenel [94], even with slight solution of the solid component in the liquid phase there is a constant interchange of atoms. The most mobile atoms on the edges and corners of the solid particles, having a large free-energy reserve, go into solution and are then reprecipitated in areas where the atoms are least mobile. Thus, there is continu-

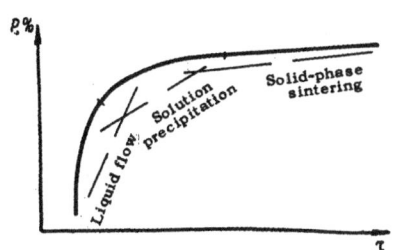

Fig. 1. Densification curve for three stages of liquid-phase sintering.

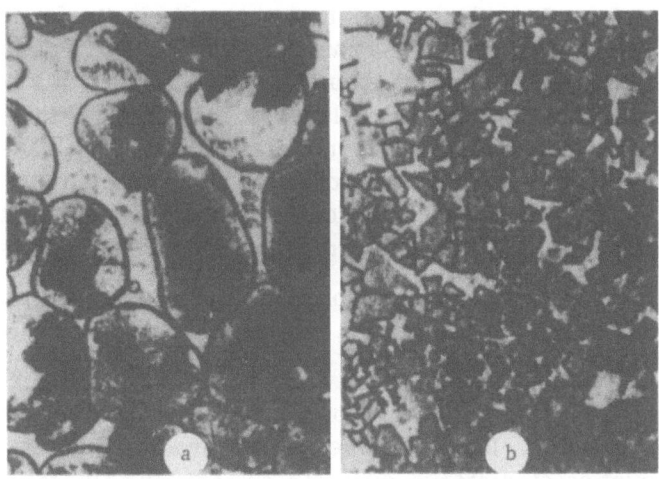

Fig. 2. Microstructures formed in liquid-phase sintering. a) Fe—Cu (× 450); b) Co—WC (× 100).

ous concentration of atoms through the liquid phase to points in contact with the solid particles, accompanied by intensive shrinkage during sintering. The growth of the particles and their shape were explained by Bal'shin and Lenel, as in [105], by the solution—reprecipitation process.

Bal'shin explained the high degree of densification during liquid-phase sintering as follows. On the appearance of the liquid phase the number of mobile atoms of the basic component increases sharply and then slowly decreases during isothermal holding. In addition, owing to the high fluidity of the liquid phase (and consequently the mobility of solute atoms), the filling of almost inaccessible pores and capillaries in the material is facilitated and accelerated. Kingery gave a different explanation of the densification mechanism [85]. According to the semiquantitative theory of liquid-phase sintering that he formulated, the substance is reprecipitated not on the contact areas but on the free surfaces of the solid particles. Kingery explained this by the fact that a layer of liquid on the contact areas creates high capillary pressures, due to which the solubility of the solid phase increases greatly in these sections and the substance is transferred from the contact areas to the free surfaces of the solid particles. As a result, the distance between the centers of the particles is reduced and the material being sintered is condensed.

Qualitative descriptions of liquid-phase sintering processes generalized from experimental data were given in [59, 74, 95]. According to these data, there are three stages of densification corresponding to three mechanisms during sintering: 1) viscous flow of the liquid — regrouping of the particles; 2) solution—precipitation; 3) solid-phase sintering with formation of a rigid skeleton.

In the first stage the liquid phase is formed, the pores are filled, and the solid particles are rearranged, resulting in close packing. In the second stage the fine particles go into solution and the substance is reprecipitated on the large particles. In the third stage the substance is slowly consolidated as the result of the solid particles growing together in accordance with the rules of solid-phase sintering. The result is that a rigid skeleton is formed in the body undergoing sintering. The predominance of one mechanism or another depends on the nature of the phases and the amount of liquid present.

Figure 1 shows a hypothetical densification curve for these three stages of liquid-phase sintering. Liquid-phase sintering results in the formation of characteristic structures consisting of evenly distributed grains of solid phase in a matrix of the crystallized liquid phase. Particles of different shapes may be formed [113]. If the surface tension at the interface between different crystallographic planes of the solid particles and the liquid phase is approximately identical (as, for example, in metals solidifying in the cubic system) then the particles have rounded shapes. If, however, the surface tension at the interface differs substantially for different crystallographic planes, then grains of prismatic shape are formed (tungsten carbide, for example). Typical microstructures of such alloys are shown in Fig. 2a, b.

In systems forming spherical particles, considerable grain growth occurs during sintering. While the grain growth is less in systems forming prismatic grains, a higher density is attained with comparatively small amounts of liquid phase.

Rearrangement of the particles is most pronounced in systems where the components are insoluble. In powder metallurgy practice such systems are frequently encountered in the production of sintered contacts (W–Cu, W–Ag, Mo–Cu systems), machine parts (Cu–C, etc.), and materials with high strength and creep resistance at high temperatures (cermets based on aluminum oxide). During regrouping, densification results from the movement of solid-phase particles under the influence of surface tension [95]. The densification is of the same character in the beginning of sintering (on formation of the liquid phase) in systems where the solid phase is soluble in the liquid phase.

This process occurs very rapidly and is the principal reason for shrinking, ensuring good densification. The volume of pores between the solid particles amounts to 25-50%, depending on the degree of packing and the particle size. With a sufficient quantity of liquid phase the theoretical density can be attained as the result of the regrouping process alone. According to Cannon and Lenel [59], this quantity of liquid phase must be at least 25% by volume. According to Kingery's calculations, if the solid-phase particles are spherical then the minimum quantity of liquid phase required for complete densification is 35 vol.%. The degree of densification decreases at smaller quantities. In this case, other sintering processes are necessary for complete densification (Fig. 3).

Solution—Precipitation is one of the processes contributing to densification at low quantities of liquid phase [59, 85]. This sintering mechanism occurs in systems whose components are fairly soluble (solid phase readily soluble in the liquid phase). The following conditions are necessary for this mechanism to occur [105]: (1) a substantial difference in the melting points of the components; (2) insolubility (or low solubility) of the low-melting metal in the high-melting metal; (3) solubility of the refractory component in the low-melting component.

As has been shown [114], the essential condition is the third. In this case the final product is heterogeneous. Densification occurs in two stages, the first involving the regrouping process and the second involving recrystallization through the liquid phase.

According to Cannon and Lenel [59], the solution-and-precipitation process cannot occur with less than 5 vol.% liquid phase. In the general case the solution-and-precipitation process results in grain growth due to the solution of fine particles in the liquid phase and reprecipitation of the substance on the large particles [105]. However, it is difficult to explain the high densification by the Price–Smithells–Williams theory, since transfer of material to large spherical particles cannot lead to complete densification without a large amount of liquid phase. Furthermore, densification continues during sintering even after the growth of the particles. Finally, despite the fact that grain growth is isometric and the samples are rapidly densified at large amounts of liquid phase, the spherical shape of the particles changes in sintered samples with a small amount of liquid phase (see Fig. 2a), the grains adjusting to each

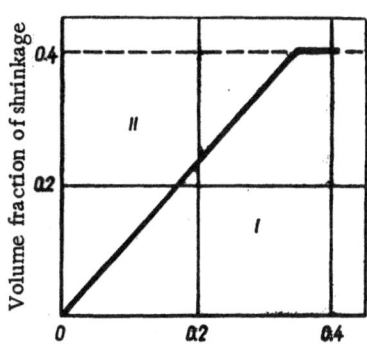

Fig. 3. Shrinkage due to regrouping (I) and to other processes (II) in relation to the amount of liquid phase (initial specimen porosity 40%).

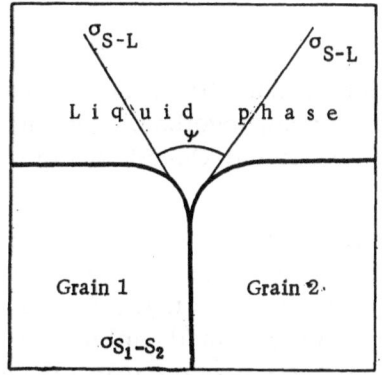

Fig. 4. Dihedral angle formed by liquid phase at the boundary of two grains of solid phase.

other's shapes, i.e., the particles acquire shapes appropriate for close packing. Consequently, partial solution and rupture of contact between particles are necessary for the particles to come closer together. Two cases of this are possible — solution over the entire surface of the particles (with subsequent reprecipitation of the material on the surface of other particles not touching them) or solution at places in contact. Obviously, the latter case is more probable. At contact points the surface tension creates high capillary pressures, favoring high solubility of the solid in the liquid phase [85]. From the resulting supersaturated liquid the solid material is then precipitated on other grains, which is responsible for the changes in the shape and the fitting together of the grains, thus promoting shrinkage.

Another important condition for the occurrence of the solution-and-precipitation process, as in the regrouping process, is the penetration of the liquid between the grains. The extent to which the liquid enters the joints between the particles depends on the dihedral angle formed by the liquid phase at the boundary with two grains of the solid phase (Fig. 4). Under equilibrium conditions this angle can be expressed as

$$\sigma_{S_1-S_2} = 2\sigma_{S-L} \cos \frac{\psi}{2}, \qquad (2)$$

where $\sigma_{S_1-S_2}$ is the interfacial energy at the boundary between two grains of the same phase, σ_{S-L} is the interfacial energy at the boundary between the solid and liquid phases, and ψ is the dihedral angle. Analysis of the equation shows that

$$\text{with } \sigma_{S-L} > \frac{1}{2}\sigma_{S_1-S_2} \quad 180° > \psi > 0;$$
$$\text{with } \sigma_{S-L} = \sigma_{S_1-S_2} \quad \psi = 120°; \qquad (3)$$
$$\text{with } \sigma_{S-L} < \frac{1}{2}\sigma_{S_1-S_2} \quad \text{no value of } \psi \text{ satisfies Eq. (2).}$$

In this case the liquid will penetrate along the boundary separating the grains of solid phase.

It is difficult to determine the angle ψ for most systems, since the values of σ_{S-L}, σ_{S-S}, and σ_{S-G} cannot be established precisely. In studying the distribution of particles in heterogeneous systems one investigates wettability and determines the angle of contact at the liquid–solid interface. When the angle of contact is near or equal to zero, the dihedral angle at the contact between two still unsintered particles is also close to zero and the liquid spreads over the surface of the grains.

Formation of Rigid Skeleton. As shown above, the extent to which the liquid flows between the solid particles depends on the value of the dihedral angle. At a certain ratio between the surface energies at the solid–solid and solid–liquid boundaries, it may turn out that the liquid does not flow into the joints between the particles. This occurs when the interfacial tension does not satisfy the relationship

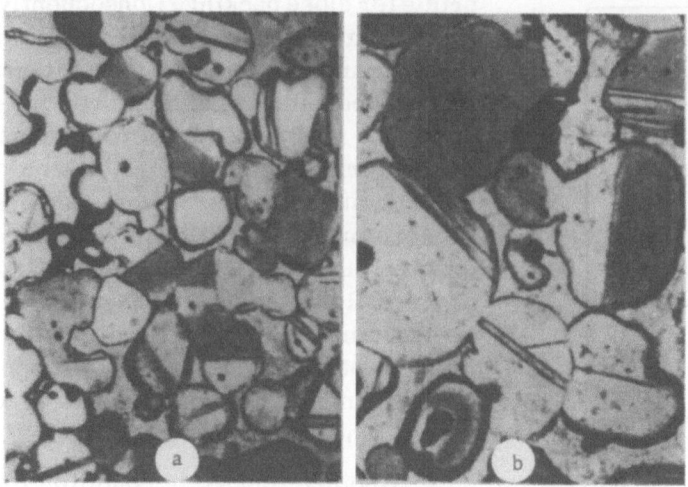

Fig. 5. Microstructure of 80 vol.% Cu−20 vol.% Bi sintered at
700°C for 1 h (a) and 24 h (b). × 300.

$$\sigma_{s-s} > 2\sigma_{s-L} \tag{4}$$

In this case neighboring particles of the solid phase may weld together. The greater the number of welded particles the more difficult is the flow of the liquid. Sometimes the number of such bonds is so great that almost all particles of the solid phase grow together, creating their own skeleton, with the liquid phase enclosed in the pores (alloys of the Cu−Bi system [59]).

In samples of the 80 vol.% Cu−20 vol.% Bi system, densification practically stops after 1 h of sintering, long before complete densification is attained. The bonding of some copper particles can be seen directly on micrographs. Further densification is very slight. It is assumed [59] that ordinary solid-phase sintering occurs in this stage, which would explain the negligible densification. After longer sintering times (up to 24 h) the grain size increases despite the fact that densification practically ceases (Fig. 5).

It has also been assumed that a rigid skeleton is formed during the sintering of carbides cemented with metals. This question was under discussion for a long time, since in carbides the surface tension at the interface with the liquid phase is not isotropic but depends on the direction of the crystal face in contact with the liquid and varies with the orientation of the grain. Therefore, the interfacial tension can differ with the conditions, which affects the results of sintering. However, as noted by Lenel, for carbides cemented with cobalt it is characteristic that the density approaches the theoretical even with small quantities of liquid phase. Such a high density can be obtained only where the particles of the solid phase become closely packed in the process of solution and precipitation, which prevents the formation of a rigid skeleton, or, if the skeleton is formed, it does so after the completion of densification. Gurland and Norton [75, 76] arrived at the same conclusion. On the basis of experiments including those on electrolytic etching and leaching, they showed that no continuous skeleton of tungsten carbide is formed as the result of sintering. Cech [60] also considers that the optimum properties of sintered carbides cannot be due to the welding of carbide grains but result from the interaction of carbide grains and the binder.

The opposite opinion is held by Dawihl and Hinnuber [67], who assume that a rigid skeleton is retained in WC−Co systems (up to 16 vol.% Co). They also found experimentally that

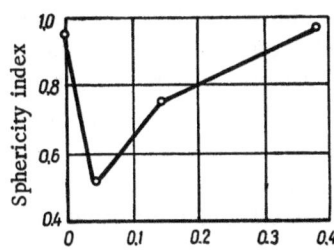

Fig. 6. Variation of sphericity index with liquid-phase concentration in Fe–Cu and Cu–Bi systems (sintered 4 h).

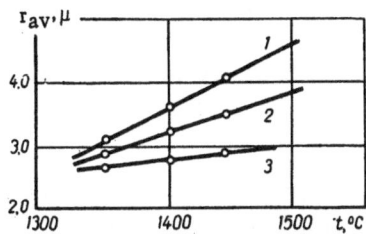

Fig. 7. Variation of average grain size of tungsten carbide in Co — WC alloys with sintering temperature and cobalt content: 1) 25 wt.% Co; 2) 16 wt.% Co; 3) 6 wt.% Co.

sintered compacts with 10 vol.% Co retain their shape and considerable strength after leaching of the binder in boiling hydrochloric acid. Bernard [52] too found a considerable number of carbide–carbide bonds in electron micrographs.

In a recent work [77] Gurland gave a detailed review of the structure and properties of cemented carbides and generalized the results obtained. Gurland takes a middle view: the structure of the carbide phase is discontinuous during the initial stages of sintering, but the number of contacts between carbide grains increases with further development of the process.

In a general case the extent to which the carbide phase is continuous depends on the composition of the sintered material, the grain size, and the sintering temperature and time. The same reasoning will, of course, apply also in the case of other alloys similar to cemented carbides (nitrides, borides, silicides).

At the present time the question is being studied in detail. A number of papers have been published concerning interfacial surface energies under various conditions as well as the values of dihedral angles and their influence on densification processes during liquid-phase sintering and on the structure of sintered bodies [61].

3. The Influence of Various Factors

on Liquid-Phase Sintering

Effect of Amount of Liquid. All the mechanisms described above usually occur in liquid-phase sintering. However, one mechanism or another may predominate, depending on the solubility of the components, the degree of wetting, and the amount of liquid phase. Kingery [86] noted that, with a large quantity of liquid, complete densification can be attained only by the process of liquid flow without changes in the shape of the grains. Kingery investigated a number of systems and found distinct differences in the microstructure of samples with different amounts of liquid phase. He calculated the sphericity index — the mean ratio of the minimum and maximum particle sizes.

The variation of the sphericity index with the amount of liquid phase in different systems is shown in Fig. 6. It can be seen from the figure that the maximum deviation occurs at small amounts of liquid phase. At such amounts of liquid phase, densification of the components requires that the shape of single particles adapt to the shape of the others in the solution-and-precipitation process. In systems in which the solution-and-precipitation mechanism is dominant the amount of liquid phase also affects grain growth. As can be seen in Fig. 7, the average grain size of tungsten carbide increases with cobalt concentration at all sintering temperatures [78].

In general, the rate and degree of densification increase with the amount of liquid phase, although in some systems considerable sample growth occurs owing to substantial solubility of the com-

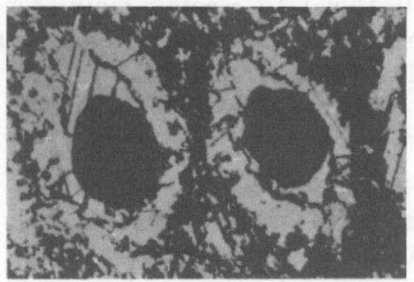

Fig. 8. Microstructure of Cu—Al alloy (10 at.% Al) sintered at 560°C for 1 h. × 450.

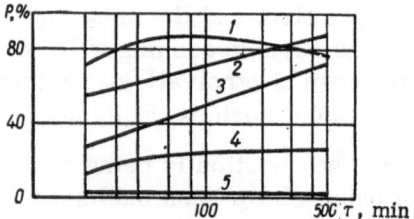

Fig. 9. Variation of densification factor for samples with 80 wt.% Fe and 20 wt.% Cu in relation to Fe particle size (sintered at 1120°C): 1) 0–7½ μ; 2) 7½–15 μ; 3) 15–30 μ; 4) 30–60 μ; 5) 44–88 μ.

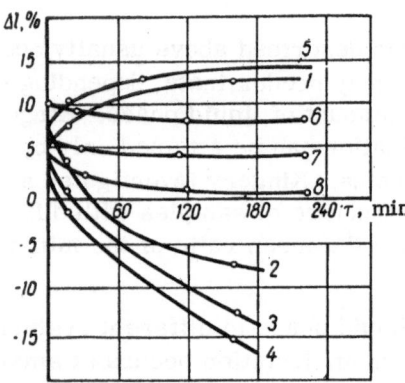

Fig. 10. Linear dimensions of pressed samples of copper with low-melting additions in relation to sintering time in hydrogen (1,2,3,4) and in vacuum of 10^{-4} mm Hg (5,6,7,8). 1,5) Cu; 2,6) Cu + 3 wt.% Sb; 3,7) Cu + 3 wt.% Pb; 4,8) Cu + 3 wt.% Sn.

ponents and the preferential diffusion of one component into the other. It has been shown [43] that, regardless of the sintering temperature, cavities are formed in aluminum as the result of aluminum diffusing into copper (Fig. 8). The cavities are structural defects and stress concentrators which affect the mechanical properties of the material. In order to eliminate the cavities, the material is re-pressed and resintered at high temperature.

Effect of Particle Size. No general principles of the effect of particle size on sintering have been established, although most investigators agree that the influence must be substantial. Kingery [85] has shown that the degree of densification during the regrouping process is inversely proportional to the particle size. In the solution-and-precipitation process the densification is inversely proportional to the particle radius to the power of $4/3$. There are no other data in the literature to confirm or contradict Kingery's findings.

During regrouping, the particle size of the solid phase has a considerable effect on densification [59]. Lenel found that marked densification occurs in samples of the W—Cu system prepared from powders with a particle size of up to 3–5 μ; in samples with larger tungsten particles, it is very small. The Fe—Cu system has been studied in more detail [59]. An increase in the densification of samples was observed when the particle size of the starting powders was small. Figure 9 shows the effect of iron particle size on the densification factor of samples with 20 wt.% Cu. Samples made from fine powders, despite less densification during compacting, exhibit better sintering and have higher mechanical properties than samples made from coarse powders.

Effect of Compacting Pressure. There are several difficulties in studying the effect of compacting pressure since, on the one hand, pressing promotes intimate contact of the particles which has a positive effect on sintering, but, on the other hand, favors the formation of "closed" pores in the sample, which sometimes has a negative effect. This is due to the fact that the gas pressure in the closed pores counteracts the capillary pressure during sintering. If the capillary pressure is higher than the gas pressure in the pores, the sample shrinks, but if the capillary pressure is lower, the sample expands. With increasing sintering time the expansion of strongly compacted samples decreases due to the compensating effect of shrinkage.

Obviously, an increase in the size of closed pores due to their coalescence reduces the gas pressure in the pores more than the capillary pressure as a consequence of the increase in the radius of curvature of the pores. In those cases where no clear effect of closed pores is manifest during sintering, the compacting pressure has different effects in different systems.

In systems of the Mo—Cu type the absolute shrinkage is inversely proportional to the compacting pressure. This can be explained by the wedging of the particles at high compacting pressures, preventing their sliding when the liquid phase occurs. This effect has not been investigated for most systems in which the solid phase is soluble in the liquid, and in which, consequently, the solution-and-precipitation process occurs during sintering. However, it is known that in such systems the compacting pressure has little effect on the final density of the sample, which is evidently due to the solubility of the material in contact zones. The effect of compacting pressure on the shrinkage during sintering of compacts from copper powders with lead, tin, antimony, and cadmium was studied in [40]. At pressures up to 900 daN/cm^2 the shrinkage varies monotonically up to a certain limit, while at pressures over 900 daN/cm^2 the samples expand, which can be explained by the occurrence of closed pores. This was confirmed in an investigation of samples of the same systems sintered in a vacuum and in hydrogen (Fig. 10).

Plate [112] has noted that the compacting pressure affects not only the shrinkage but also the initial interaction temperature of the components (in the Fe—Zn system, for example). Nevertheless, it can be concluded that in most cases an increase of compacting pressure has a positive effect on sintering — in all cases where it does not result in the formation of closed pores.

Effect of Sintering Medium. The medium is one of the major factors in sintering, its effect differing little in solid-phase and liquid-phase sintering.

In all cases the medium serves as a protection against the influence of oxygen and water vapor. The protection may be passive (inert gas) or active (hydrogen). Purified inert gas reduces the partial pressure of oxygen and water vapor, preventing oxidation of the material, but does not break down any oxide films that may exist before sintering begins. Hydrogen reduces oxide films, resulting in rapid and improved sintering, since the surface of the particles is freed of oxides and becomes rough.

Vacuum is a passive protection on one hand, since it does not reduce oxide films, but is active on the other hand, since it promotes the vaporizing of these films and the removal of adsorbed gases and gases from open pores,* thus accelerating the sintering process.

The effect of the medium on the sintering process, including liquid-phase sintering, has been studied in detail for a number of systems [49, 100]. We shall not discuss the effect of other factors on liquid-phase sintering, such as the method of preparing the powder, vaporization of the liquid phase, its viscosity, the presence of impurities, etc. There have been few special studies of these questions and it is impossible to say anything definite even with regard to individual systems.

4. Elements of the Theory of Liquid-Phase Sintering

The concepts presented above are basically a qualitative, not quantitative, description of liquid-phase sintering. The first attempt at a quantitative description of liquid-phase sintering and the effect of different factors on the process was made by Kingery [85]. Kingery assumes that complete densification of a powder compact by liquid-phase sintering requires the presence of some specific amount of liquid in which the solid phase is soluble (he considers only

---

*The attenuating effect of trapped gases in closed pores on shrinkage has been examined in detail by Cech.

systems in which the solid phase is wetted completely by the liquid). Under these conditions shrinkage depends on three processes. The first of these — regrouping of the solid particles — depends on the capillary pressure of the liquid formed, which tends to redistribute the particles in such a way that the residual porosity is minimal. This occurs by means of the particles sliding with respect to each other, which in general corresponds to viscous flow of the material. Assuming that densification in this stage depends entirely on viscous flow, Kingery gives a kinetic equation of shrinkage

$$\frac{\Delta l}{l_0} = \frac{1}{3} \cdot \frac{\Delta V}{V} = A\tau^{1+x},$$ (5)

where the exponent $(1 + x)$ is greater than unity, since the pore size decreases and the driving forces of the process increase during sintering.

Theoretical studies of the regrouping process have also been made by Cech [62], who also considers that the main driving force of sintering is capillary pressure, but distinguishes two stages of sintering — before and after the closing of pores in the body undergoing sintering.

If it is assumed that viscous flow under the influence of the surface tension of the liquid phase occurs in the first stage, then the equation of strain can be written as

$$\frac{dl}{l_0} = \frac{1}{\eta} \, vd\tau,$$ (6)

where η is the viscosity coefficient of the substance, l_0 is the original size of the body, and v is the stress in the direction of the acting force.

In place of v in Eq. (6), Cech substitutes the pressure difference causing viscous flow

$$\Delta p = p - p_v = \frac{2\sigma}{r} - p_v,$$ (7)

where p is the pressure in a spherical void of radius r and p_v is the counterpressure of the gas in the spherical void. Then

$$\frac{dl}{l_0} = \frac{1}{\eta} \cdot \frac{2\sigma}{r} \, d\tau - \frac{p_v}{\eta} \, d\tau.$$ (8)

For sintering in a vacuum it can be assumed that $p_v = 0$, and then the second term in the equation vanishes. For the closing of a void in a sample during viscous flow, Frenkel' derived the equation

$$\frac{dr}{d\tau} = - \frac{3}{4} \frac{\sigma}{\eta_L}$$ (9)

(where η_l is the viscosity coefficient of the liquid), from which it follows that

$$r = - \frac{3}{4} \cdot \frac{\sigma}{\eta_L} \, \tau.$$ (9')

Substituting this relationship into (8), we obtain

$$dl = - k_1 l_0 \frac{d\tau}{\tau},$$

where

$$k_1 = \frac{8\eta_L}{3\eta}.$$ (10)

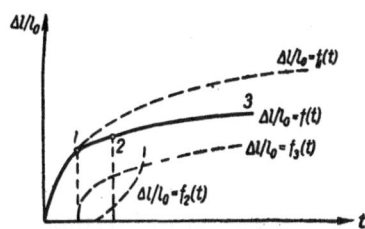

Fig. 11. Sintering curves in the region of viscous flow. The shrinkage in different sections of the curve is determined by the following functions: sections 0-1) $\Delta l/l_0 = f_1(\tau) = k_1'(\log \tau - \log \tau_0)$; section 1-2) $\Delta l/l_0 = f_1(\tau) - f_2(\tau) = k_1'(\log \tau - \log \tau_0) - k_2(\tau^4 - \tau_k^4)$; section 2-3) $\Delta l/l_0 = (1-2) + f_3(\tau) = (1-2) + k_3\tau^{1/3}$.

Thus, the rate of shrinkage decreases with time according to

$$\frac{dl}{d\tau} = -k_1 l_0 \frac{1}{\tau}. \tag{11}$$

Integrating from l_0 to l (l_0 corresponds to the initial pores r_0 at τ_0), we obtain

$$\frac{l_0 - l}{l_0} = \frac{\Delta l}{l_0} = k_1'(\log \tau - \log \tau_0). \tag{12}$$

This relationship is illustrated in Fig. 11.

According to (8), after filling of pores the shrinkage in a protective gas should be described by

$$\frac{\Delta l}{l_0} = k_1'(\log \tau - \log \tau_0) - k_2(\tau^4 - \tau_k^4), \tag{13}$$

where the second term, designated $f_2(\tau)$ in Fig. 11, is calculated by substituting the expression $\frac{4}{3}p\pi r^3$ (from the relationship $p_0V_0 = pV = \frac{4}{3}p\pi r^3$) and r is taken from Eq. (9'); k_1' and k_2 are proportionality factors.

Later, the sintering mechanism of solid particles not in contact will be determined by the diffusion of the compressed gas in the pores. In this case the gas concentration gradient depends on the external pressure and the pressure in the pores.

Using Fick's law for the diffusing substance and assuming that the external pressure is small by comparison with the pressure in the pores, one can substitute the concentration gradient in the following form:

$$\frac{dc}{dx} = \frac{2\sigma}{r} \cdot \frac{1}{x(1 - 2rn)}, \tag{14}$$

where n is the number of pores in length x, and r is the pore radius.

Then the quantity of gas diffusing from a sphere of radius x is

$$dp = -D4\pi x^2 \frac{2\sigma}{r} \cdot \frac{1}{x(1 - 2rn)} d\tau, \tag{15}$$

where D is the diffusion coefficient.

At $dp \approx dV$ (dV is the reduction in the volume of pores):

$$dp \approx \frac{3V_0}{r_0} dr \approx -4D\pi x^2 \frac{2\sigma}{r} \cdot \frac{1}{x(1 - 2rn)} d\tau. \tag{16}$$

Performing suitable transformations and integrating from r_0 to r (r_0 was calculated for the case of the closing of pores at the corresponding sintering time τ_0, while r corresponds to time τ), we obtain

$$\tau_0 - \tau = \frac{1}{D} \cdot \frac{V_0}{r_0} \cdot \frac{N_1}{4\pi\sigma}(r^3 - r_0^3), \tag{17}$$

where N_1 is the number of pores per unit length.

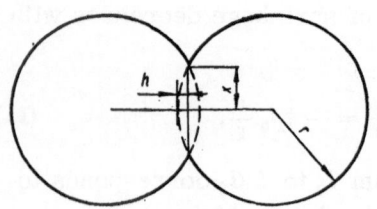

Fig. 12. Diagram of the drawing together of spherical particles resulting from solution of the substance at the contact surface.

If sintering proceeds up to r = 0, then from (17) we obtain

$$\frac{\Delta l}{l_0} = k_3 \tau^{1/3},$$ (18)

where k_3 is a constant.

Figure 11 shows the resultant curve for these processes in the form of $\Delta l/l_0 = f(\tau)$.

Kingery's theory finds confirmation in [93], in which the initial stage of isothermal shrinkage is described by

$$P = A\tau^n,$$ (19)

where P is the densification parameter,* τ is time, A is a constant, and n is equal to 0.3–0.5. For good agreement between theoretical and experimental results, a correction factor is introduced into Eq. (19): $(1 - P)^i$, the so-called "encounter factor" (i is selected arbitrarily).

The encounter factor accounts for the processes determining the shrinkage of samples — the meeting of neighboring particles, the decrease in the particle surface or volume, and the changes in the pore size and shape. Thus, Eq. (19) is written

$$P = A(1 - P)^i \tau^n.$$ (19')

It was found that n does not change on reduction in the volume of pores if i = 1. Then,

$$V = \frac{P}{1 - P} = A\tau^n; \quad \frac{dV}{d\tau} = nA\tau^{n-1}.$$ (20)

From (20) it follows that

$$\tau = \left(\frac{V}{A}\right)^{1/n},$$

and, consequently,

$$\frac{dV}{d\tau} = nA\left(\frac{V}{A}\right)^{1 - \frac{1}{n}}.$$ (21)

Designating $nA^{1/n} = D$, we obtain

$$\frac{dV}{d\tau} = DV^{\left(1 - \frac{1}{n}\right)}.$$ (22)

The effect of temperature is taken into account by the relationship

$$D = D_0 e^{-\frac{Q}{RT}},$$ (23)

---

*The densification parameter is determined as

$$P = \frac{d_s - d_{ns}}{d_{theor} - d_{ns}},$$

where d_s is the density of the sintered sample, d_{ns} is the density of the unsintered sample, and d_{theor} is the theoretical density.

where Q is the activation energy, R is the gas constant, D is the apparent diffusion coefficient, and T is absolute temperature.

Experimental verification of Eq. (22) for sintering compacts from a mixture of iron and copper powders showed that it holds true only within a limited range of temperatures and times.

At the end of the first stage of the process, close packing of the particles is ensured by the capillary pressure and, according to Kingery, is completed when the capillary pressure drawing the particles together is balanced by the compressive stresses in the contact area.

The stressed state of the contacts causes a chemical potential gradient and increases the solubility of the substance in the contact area in accordance with the relationships

$$\mu - \mu_0 = RT \ln \frac{a}{a_0} = \Delta p V_0, \tag{24}$$

$$\ln \frac{a}{a_0} = \frac{k 2\sigma_L V_0}{r_p RT}, \tag{25}$$

where μ and a are the chemical potential and activity of the refractory component in the liquid phase within the contact zone, μ_0 and a_0 are the chemical potential and activity of the refractory component in the liquid around the free surface, σ_L is the surface tension of the liquid, V_0 is the molecular volume, r_p is the radius of the pore, T is the absolute temperature, and k is a constant relating the maximum pressure on the contact surface to the total hydrostatic pressure.

The dissolved substance diffuses in the liquid phase and is precipitated on the free surfaces of solid particles. As a consequence, the particles will be drawn together under the influence of capillary pressure, leading to further shrinkage. For a quantitative description of this process it is necessary to know the elastic forces generated in the contact area and also the surface energies and diffusion constants. It is difficult to determine these constants under the complex conditions of liquid-phase sintering. However, the functional variation of shrinkage with the sintering time and temperature and particle size can be approximated.

Let us consider the sintering of two spherical particles of radius r brought into contact under the influence of surface tension (Fig. 12). Let us assume that solution occurs in the contact zone. Then, during the solution of each sphere, a round contact zone of radius x will occur at a distance h along the line connecting the centers of the particles.

The volume of material removed from one sphere will be

$$V = \frac{\pi x^2 h}{2}. \tag{26}$$

Since

$$x^2 = 2rh - h^2,$$

one can neglect h^2 when $h \ll x$. Then,

$$x^2 = 2rh, \quad V = \pi r h^2, \quad dV = 2\pi r h \, dh. \tag{27}$$

It can be assumed that the rate at which the substance is removed from the contact area depends on the diffusion flux from the contact surface to the surrounding region. If we use the analogy between the diffusion flux of the substance and the heat flux from the center to a cooled surface, then for a cylindrical solid we can write

$$J = 4\pi D (C - C_0), \tag{28}$$

where J is the diffusion flux across the boundary per unit thickness.

For a boundary thickness δ

$$\frac{dV}{dt} = \delta J = 4\pi\delta D\,(C - C_0).$$

(29)

Let us assume that the pressure at the contact surface is inversely proportional to the ratio of the contact surface to the sintering surface of the particles. Then

$$\Delta p = \frac{k_2 p_0}{(x^2/r^2)} = \frac{2k_2\sigma_L\cdot r^2}{r_p\,2rh} = \frac{k_2\sigma_L}{k_1 h},$$

(30)

where r_p is the pore radius, k_1 is the ratio of the pore radius to the particle radius, and k_2 is a proportionality factor.

Combining Eqs. (24) and (30), we have

$$(C - C_0) = C_0\Big[\exp\Big(\frac{k_2\sigma_L V_0}{k_1 hRT}\Big) - 1\Big].$$

(31)

Let us assume that the value of k_1 in Eq. (31) remains constant during sintering. Since the reduction in the volume of the body matches the reduction in the volume of pores, then, taking Eq. (26) into account, we can write

$$\Delta V = \frac{4}{3}\pi\,(r_{p_0}^3 - r_p^3) = \frac{\pi x^4}{16r}$$

(32)

and

$$r_p = \Big(r_{p_0}^3 - \frac{3x^4}{64r}\Big)^{1/3}.$$

(33)

The reduction in the volume of spherical particles is equal to the volume of the substance transported across the perimeter of the contact surface:

$$\frac{dV}{d\tau} = 4\pi\delta D\,(C - C_0) = 4\pi\delta DC_0\Big[\exp\Big(\frac{k_2\sigma_L V_0}{k_1 hRT}\Big) - 1\Big] = \frac{2\pi rhdh}{d\tau}.$$

(34)

If we expand the exponent in a series and consider only the first terms of the expansion, we obtain

$$h^2 dh = \Big(\frac{2k_2\delta DC_0\sigma_L V_0}{k_1 RT}\Big)\,r^{-1}d\tau.$$

(35)

Integrating Eq. (35), we find

$$h = \Big(\frac{6k_2\delta C_0 D\sigma_L V_0}{k_1 RT}\Big)^{1/3}\,r^{-1/3}\tau^{1/3}.$$

(36)

Since $h/r = \Delta l/l_0$, the shrinkage will be expressed by

$$\frac{\Delta l}{l_0} = \frac{1}{3}\cdot\frac{\Delta V}{V} = \Big(\frac{6k_2\delta DC_0\sigma_L V_0}{k_1 RT}\Big)^{1/3}\,r^{-4/3}\tau^{1/3}.$$

(37)

A similar relationship can be derived, assuming that the transport of substance occurs under kinetic conditions at the interface. In that case,

$$\frac{\Delta l}{l_0} = \frac{1}{3}\cdot\frac{\Delta V}{V} = \Big(\frac{2k_3 k_2\sigma_L C_0 V_0}{k_1 RT}\Big)^{1/3}\,r^{-1}\tau^{1/3},$$

(38)

where k_3 is a proportionality factor.

A similar relationship can be obtained for the model of sintering prismatic particles. However, since prismatic particles may be arranged in any pattern, the geometric relationship is not as simple as for contacting spheres. Therefore, it is very difficult to find a functional relationship in this case. Nevertheless, for the early stages of the process, where maximum shrinkage occurs in the contacting zones and the rate of material transport depends on the diffusional flux, one can obtain $\Delta l / l_0 \sim \tau^{1/5}$ and, for the reaction at the interphase boundary, $\Delta l / l_0 \sim \tau^{1/3}$.

The question of which is the controlling mechanism is of major practical importance.

Skolnick [115] investigated the solution kinetics of tungsten carbide (WC) in liquid cobalt and found a very low solution rate (a value two orders below that calculated for the diffusion-controlled process), the solution rate being independent of the rate of revolution of the sample. In this case, the activation energy of the process was high and amounted to 175 ± 45 kcal/mole. These data indicate that the solution process operates under kinetic, rather than diffusion, conditions. These observations have not yet been confirmed or disproved in studies of other systems.

It should be noted that the current shrinkage-mechanism theory does not take into consideration the quantitative effect of powder characteristics, interdiffusion of the components, the amount of liquid phase, etc. Data on the solubility of refractory components in the liquid phase and on diffusion under these conditions are also almost completely lacking. This shows the necessity of further studies in order to formulate a more complete theory of liquid-phase sintering and to use it for solving practical problems.

CHAPTER 2

CAPILLARY PHENOMENA AND WETTABILITY IN LIQUID-PHASE SINTERING PROCESSES

1. The Substance Being Sintered

Analyzed as a Dispersed Capillary System

The formation of a powder metallurgical body during liquid-phase sintering occurs in a system consisting of solid, liquid, and gaseous phases. Since the particle size of the solid phase is usually small (of the order of $1\,\mu$) and the pore size is of the same order, the sintered body is a capillary system. The behavior of such a system with highly developed surfaces, the distribution of phases within it, and consequently the properties of the body itself, will depend to a considerable extent on the properties of the boundaries between the phases within the system (the solid—liquid, solid—gas, and liquid—gas boundaries).

The field of forces acting on the dispersed particles in such a system depends on the surface energies of the interphase boundaries. Therefore, the kinetics of sintering and densification will also depend to a considerable extent on the capillary properties of the system.

Let us examine these assumptions in more detail.

First of all, let us consider the characteristics of the solid-phase composite — a conglomerate of fine solid-phase particles. In the simplest case, where the body is formed of spherical particles, the porosity of the body, i.e., the ratio of the volume of pores to the volume of the whole material, depends on the packing system of the particles. Cubic packing results in the lowest density (Fig. 13a), the porosity amounting to 47.64% irrespective of the particle size. Each elementary pore has the form of an octahedron, the faces of which are concave spherical surfaces (Fig. 14a). The highest density results from hexagonal or rhombohedral packing (Fig. 13b), the porosity amounting to 25.95%. The elementary pore has either a tetrahedral or rhombohedral form, the porosity amounting to 7.36 and 18.58%, respectively (Fig. 14b).

Real porous bodies, even those composed of spherical particles, are not usually densely packed. In an unpressed substance the packing is approximately cubic (with spherical as well as nonspherical particles), and the porosity of such a substance is around 40% [86].

The behavior of the liquid at the contact with the solid phase is characterized by the angle of contact θ at the solid—liquid—gas boundary. The height of the capillary rise of the liquid in a vertical capillary of radius r is determined by

$$H = \frac{2\sigma_L \cos\theta}{rg(\rho_L - \rho_V)},\tag{39}$$

where ρ_L and ρ_V are the densities of the liquid and vapor, respectively; g is the acceleration due to gravity.

16

Fig. 13. Packing of spherical particles. a) Cubic; b) rhombohedral.

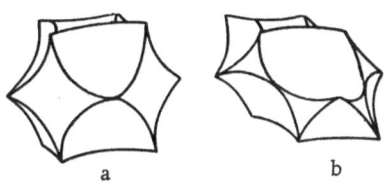

Fig. 14. Shape of pores in two types of packing of spherical particles. a) Cubic; b) rhombohedral.

It should be emphasized that at $\theta < 90°$ the liquid tends to be drawn into the capillary; at $\theta > 90°$, the liquid is forced out of the capillary.

The behavior of a liquid in different capillaries can be characterized by the capillary potential, defined as the "potential energy of the field of capillary forces per unit mass of liquid" (ergs/g) [29].

By analogy with the potential of the field of gravity, the capillary potential for a cylindrical capillary is

$$\psi_c = gh = \frac{2\sigma_L \cos \theta}{r\rho_L} \qquad (40)$$

at $\rho_L \gg \rho_V$. Thus, a wetting liquid tends to occupy a position with the highest capillary potential, and a nonwetting liquid a position with the lowest potential. As the result of this a wetting liquid spontaneously moves from a wide into a narrow capillary, the system tending toward a lower energy state.

The properties of the interfaces in a capillary system and the wettability of the solid phase with the liquid determine many characteristics of the system that are important in the liquid-phase sintering process and in the production of sintered metal parts with the desired properties.

Area of Contact Between the Liquid and Solid Phases. The area of contact in the system of solid particles under consideration is directly proportional to the extent of wetting of the particles by the liquid phase. With decreasing angle of contact (angles different from zero) the area of contact between the liquid and solid surfaces increases. When

$$\sigma_L + \sigma_{s-L} < \sigma_{s,} \qquad (41)$$

the area of contact must be a maximum, i.e., the surface of the solid particles must be completely covered with a film of liquid.

Distribution of Phases in the Capillary System. In a system of solid particles there are capillary pores and gaps of different sizes and shapes, and therefore the value of the capillary potential differs at different points in the system, i.e., the energy state of the liquid differs at different places in the dispersed system. The liquid will tend to occupy positions ensuring the minimum energy of the system as a whole.

In a body formed from solid spherical particles of identical size, it is possible to distinguish certain characteristic areas over which the liquid may be distributed:

a) points of contact between solid particles;

b) cups or nodoids of liquid in the vicinity of the contact areas;

c) pores (tetrahedral, octahedral, etc., depending on the type of packing).

When the quantity of liquid is insufficient for the complete filling of all pores, its distribution in a porous body will vary depending on the magnitude and ratio of surface energies at the phase boundaries. If the liquid readily wets the solid particles and the dihedral angle at the contact is equal to zero, and consequently,

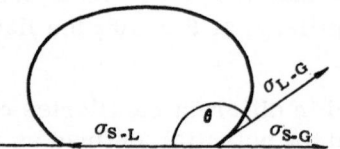

Fig. 15. Diagram of equilibrium
drop of liquid on solid surface.

$$\sigma_{S\text{-}L} < \frac{1}{2}\sigma_{S\text{-}S'} \qquad (42)$$

then the liquid will penetrate to the points of contact.
Under these conditions, too, cups of liquid will occur
(noted in point b above). The pore volume, however, will
be free from liquid (areas noted in point c above).

If the angle of contact $\theta < 90°$ but condition (42) is
not fulfilled, the liquid will not penetrate into the contact
areas and the contact between solid particles will not be disrupted. At a suitable quantity of
liquid, the pores will also be empty, and only cups or nodoids will be filled.

At $\theta > 90°$ only the pores will be filled (point c above). In this case the system is un-
stable and the liquid tends to separate from the porous solid substance.*

Capillary Forces. A wetting liquid preferentially forms cups. Capillary forces
act on the curved liquid–gas surfaces, meniscuses, and on the three-phase (liquid–solid–
gas) contact lines, pulling together or pushing apart neighboring particles. The magnitude of
the capillary forces can vary depending on the size of the particles in the system, generating
stresses in the substance being sintered that reach tens and hundreds of daN/cm². The compressive forces are the driving force of the sintering process and promote densification of the
compact. The magnitude of the compressive forces depends to a considerable extent on the
angle of contact, the amount of liquid, and the size and shape of the solid particles.

2. Wetting of Solids with Liquid Metals

The properties and manufacturing technology of cermets and the formation of a strong
bond between phases, particularly phases with different physicochemical natures (especially
metals in contact with nonmetals), are determined to a great extent by the wetting of solid particles with liquid metal.

A liquid will spread over a smooth solid surface if the free energy of the system decreases when the area of contact between the liquid and solid surface increases:

$$dF < 0.$$

If σ_L, $\sigma_{S\text{-}L}$, and σ_S are the surface energies, while A_L, $A_{S\text{-}L}$, and A_S are, respectively, the
areas of the liquid–gas, liquid–solid, and solid–gas interfaces, then

$$dF = dA_L\sigma_L + dA_{S\text{-}L}\,\sigma_{S\text{-}L} + dA_S\sigma_{S'}$$

Taking into account that the changes in area during spreading are related as

$$dA_L = dA_{S\text{-}L} = -dA_S,$$

we obtain the condition for spreading

$$\sigma_L + \sigma_{S\text{-}L'} - \sigma_S < 0.$$

If $dF > 0$, the liquid joins the surface of the solid at an angle (Fig. 15) which can be found
from the condition of equilibrium of the vectors of the surface forces [6]:

---

*In sintering practice this phenomenon is frequently observed in the form of liquid drops exud-
ing from the pores of compacts.

$$\cos \theta = \frac{\sigma_S - \sigma_{S-L'}}{\sigma_L}. \qquad (43)$$

By simple reasoning we obtain

$$W_{ad} = \sigma_S + \sigma_L - \sigma_{S-L'} \qquad (44)$$

where W_{ad} is the work of adhesion, i.e., the work of detaching the liquid from the solid surface.

Thus, the angle of contact depends on the relative magnitude of the free surface energies at the three interphase boundaries. An analysis of the effect of interphase energies on the angle of contact was given in [15]. Here, it should be emphasized that the claim to be found in the literature that substances with large surface energies σ_S exhibit better wettability is, generally speaking, incorrect. In fact, if the value of σ_S varies at constant values of σ_{S-L} and σ_L, the wettability increases with increasing surface tension of the solid phase. This experimental condition can be realized, for example, when some changes occur on the part of the solid surface not wetted with liquid (for example, on desorption of foreign molecules, inducing an increase of σ_S).

However, when one solid substance with σ_{S_1} is replaced by another with σ_{S_2}, a change in the surface tension at the solid–gas interface is inevitably accompanied by a change in the tension at the solid–liquid interface, since

$$\sigma_{S-L} = \sigma_S + \sigma_L - W_{ad}. \qquad (45)$$

The angle of contact is determined by the difference $(\sigma_S - \sigma_{S-L})$ and not by the absolute values of σ_S and σ_{S-L} independently.

Substituting Eq. (45) into (43), we obtain

$$\cos \theta = \frac{W_{ad} - \sigma_L}{\sigma_L}, \qquad (46)$$

i.e., the angle of contact depends on the properties of the liquid (σ_L) and the intensity of the interaction between the liquid and the solid (W_{ad}). However, the form of the relationship $\theta = f(\sigma_S)$ is not determined by Eqs. (44)–(46).

The surface tension of pure liquid metals has been studied in some detail [44]. A characteristic feature of the free surface energy of metals is the extraordinary dependence of σ on even very small amounts of impurities. Slight amounts of surface-active elements in a liquid metal can reduce the surface tension to one-half or less, and this must be taken into consideration in liquid-phase sintering.

A fairly high degree of wetting of a solid surface by liquid metal can occur during intensive chemical reaction at the interface. This is a general rule of the wetting of solids by various liquid metals.

Depending on the nature of the solid and liquid phases, this reaction may differ both in character (formation of an intermediate phase, solution, or diffusion in the solid) and in mechanism (nature and structure of the intermediate product, solution mechanism). Therefore, in considering data on the contact properties of liquid metals in relation to different solids, it is expedient to classify the latter as follows: refractory oxides of metals; carbon materials (graphite, diamond); compounds of metals with carbon, boron, and nitrogen – carbides, borides, and nitrides; hard metals.

Metal—Oxide Systems. Refractory oxide materials generally exhibit only slight wetting by liquid metals. For such metals as mercury, tin, lead, silver, copper, nickel, cobalt, and iron the angle of contact with refractory oxides is 120-150° [15]. However, the very first qualitative observations of metal—oxide contact revealed that some metals (titanium, zirconium, cerium, lanthanum, aluminum) intensively wet oxide surfaces. In studies of the wettability of oxides with liquid metals a number of significant facts have been established:

1) The wettability of an oxide increases as the affinity of the wetting liquid metal for oxygen increases;

2) the wettability of an oxide decreases with increasing free energy of formation of the oxide, i.e., with increasing energy binding the oxygen in the oxide;

3) in studies of the reaction products of an oxide and a metal, oxide phases (new oxides, spinels, silicates) are usually found.

Intermetallic compounds are as a rule only rarely encountered. It should also be taken into consideration that the surface of refractory oxides is formed by the large oxygen anions; the cations, being smaller in size, are as it were shielded by the anions. According to Weyl [122] the cations can even be forced from the surface into the oxide.

On the basis of the data obtained, it can be assumed that the reaction of a liquid metal with an oxide is determined by the reaction of the metal with the oxygen in the oxide. The reaction for divalent metals can be described by [15]:

$$Me'' + Me'O \rightleftarrows Me' + Me''O,$$
$$\Delta F = \Delta F'' - \Delta F',$$

where $\Delta F''$ and $\Delta F'$ are the free energies of oxidation of the liquid metal and the metal forming the solid oxide, respectively.

Thus, a solid oxide is best wetted by a metal that is active with respect to oxygen; in turn, this metal wets best an oxide having a lower energy of formation ΔF.

Metal—Graphite (Diamond) System. In metal—graphite (diamond) systems a wettability that is at all appreciable can be expected when the metal vigorously reacts chemically with carbon. Strong bonds of metal-like character are formed with carbon by all transition metals, i.e., elements with defective d- or f-shells. Fairly stable compounds with an ionic bond component are formed with carbon by strongly electropositive — rare earth and alkaline — elements. There is also a number of elements (aluminum, silicon, boron, etc.) forming covalent bonds with carbon. The above-mentioned, carbide-forming elements, both in the pure form and as small additions to other inactive metals, would be expected to adhere strongly to the graphite and diamond surfaces and to wet them well.

The nontransition metals, which do not react with carbon, would not be expected to wet the surface of diamond or graphite. In a study of the wetting of graphite and diamond with liquid metals (Cu, Sn, Ag, Au, Ge, In, Sb, Pb, Bi, Ga) it was found that they do not wet diamond and graphite and that their angle of contact with the surface is very large [33-35]. The transition metals (Ti, Cr, Zr, Mn, Nb, V, etc.) are strong adhesion-active elements that sharply increase the wetting of graphite and diamond when added to the inactive nontransition metals.

In certain cases, during the contact of an alloy containing an adhesion-active component, a new phase is observed at the graphite or diamond interface [33], which is a carbide of the adhesion-active metal. These observations confirm the chemical mechanism of the wetting process in systems with high surface energies. Data on wetting confirm the hypothesis that the adhesion activity of an element toward graphite and diamond increases with increasing defectiveness of the d-shell of the metal.

In comparing the wettability of diamond and graphite with different metals it is possible to use as an assessment criterion the value of the free energy of carbide formation or carbon solution in the metal.

Metal—Carbide, Metal—Boride, Metal—Nitride Systems. It has been found that the transition metals best wet the surfaces of carbides, borides, and other phases and spread over them. As a rule, the nontransition metals wet these phases less well [15, 16], i.e., the rule obtained for graphite holds true qualitatively also for the metal-like phases. Evidently, the metal-carbon (and, of course, metal—boron for borides and metal—nitrogen for nitrides) reaction plays a substantial role in the wetting of carbide phases. At the same time, the reaction of many metals in wetting metal-like compounds (carbides, borides) is more intense than in wetting pure carbon in the form of graphite or diamond. For example, the angle of contact of pure nickel on graphite is around 60°, while on carbides of titanium, chromium, and many others the angle of contact is close to zero. The angle of contact of copper on graphite is very large — 140-150° — while on chromium carbide it is 50-60°. This means that the metal atoms in metal-like phases also react with the metal of the liquid phase or that the carbon in the lattice of the solid phase is in such a state (metallic) that it can easily react with the liquid metal. Eremenko and Naidich [16] found that the lower the strength of the bond between the atoms of a metal and boron in a wetted compound, the higher the wettability of the boride. For example, titanium boride is wetted less well than molybdenum and chromium borides.

Liquid Metal—Solid Metal System. The surfaces of metals would be expected to be easily wetted with liquid metals. Investigations have shown that, at angles of contact of more than 90° (in the Fe—Pb and W—Sn systems, for example), lack of wetting is due to contamination of the contacting surfaces. In these systems, spreading of a liquid metal over a solid phase can be induced by special heat treatment of the surface, and for this reason cleaning methods, fluxes, etc., are of major practical importance.

Thus, reaction between metals is sufficient to ensure a high degree of wetting. The character and intensity of reaction between a liquid and a solid metal are also important factors. Thus, an oxide film on the contact surface readily inhibits wetting in weakly reacting systems (the term "reaction" referring to the formation of a solution or new phases), while on the other hand an intense reaction at the interface can disrupt an existing contaminant (oxide, carbide) film.

****

It follows from the above that study of contact reactions (wetting, adhesion) in liquid metal — refractory nonmetallic material or liquid metal—solid metal systems makes it possible to determine the nature and mechanism of wetting at a liquid metal—solid phase interface.

At the present time theoretical concepts have been formulated permitting at least qualitative predictions of the behavior of metals in contact with a number of refractory materials. This opens up the possibility of controlling rationally the wetting and adhesion processes and their use in liquid-phase sintering.

CHAPTER 3

METHODS OF INVESTIGATING
LIQUID-PHASE SINTERING

1. Measurements of Shrinkage. Dilatometric
Studies of Densification During Sintering

Sintering produces various changes in the sample, which can be divided into three groups [60]. The first group includes changes in volume and in the physical and mechanical properties of the sintered material (density, electrical conductivity, magnetic susceptibility, hardness, modulus of elasticity, etc.). The second group includes phase changes, the appearance or disappearance of liquid phase, and the formation of intermetallic compounds and solid solutions. The third group includes changes in the sizes of the solid complexes within the phases (the grain sizes of the structural components).

Due to the variety of properties changing during sintering one can study the process by different methods. The sintering process is frequently studied by the dilatometric method [31, 63, etc.], since the shrinkage (i.e., changes in the linear dimensions and volume of the sample during sintering) is one of the basic indicators of the process. The sintering process is also often investigated by studying the electrical conductivity or electrical resistivity of the sintered body [19, 107]. The sintering process has occasionally been followed by studying the change in hardness [75], bending strength [75], etc. Usually the basic method of investigation is supplemented with others: determination of the density, examination of the microstructure, measurement of the grain size, determination of the heat content in the process of sintering, x-ray structural analysis, etc. Only by studying a group of properties during sintering can one determine the true sintering mechanism and clarify other questions of interest.

Let us examine briefly the advantages and disadvantages of the various methods of investigation which can be used to study liquid-phase sintering processes.

Dilatometric Method. The most extensive use of this method has been made by Raub and Plate [106, 111], Duwer and Martens [68], and Goncharova [8]. Using the dilatometric method, Raub and Plate [111] investigated a large number of two-component metallic systems and determined the type and character of dilatometric curve (elongation as a function of time) for systems in which no interactions occur and for systems in which chemical reactions occur with formation of intermetallic compounds, etc.

Dilatometric studies of powder metallurgical systems have also been made by Cech and others [63, 98]. This method makes it possible to trace the densification process directly during sintering. It is particularly important in liquid-phase sintering, since the appearance of the liquid phase sharply changes the rate of the process, which is recorded on the dilatometric curves. Dilatometric measurements make it possible to establish certain principles of the sintering process and provide information on the rate of the process, the diffusion interaction of phases, and volume and surface flows of the substance, i.e., all the phenomena which are responsible for volume changes of the material during sintering.

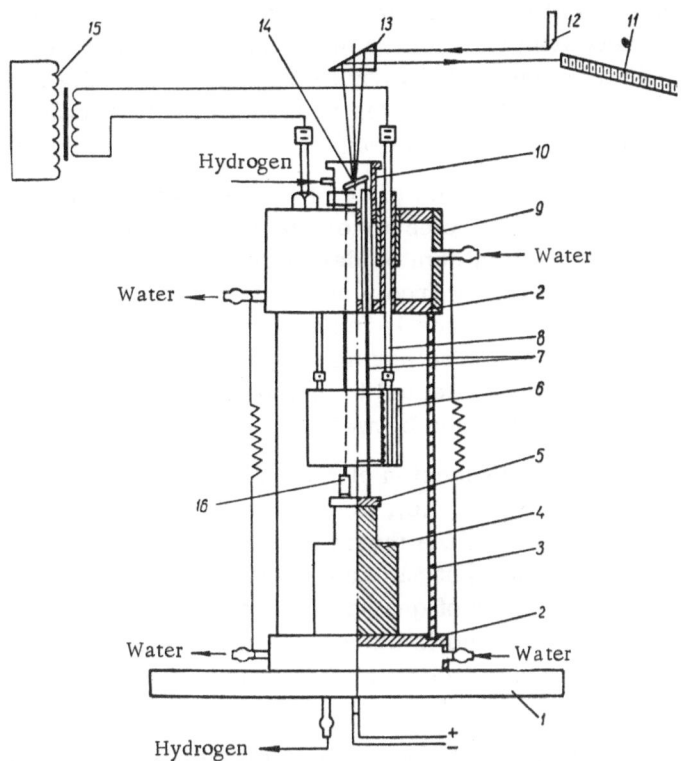

Fig. 16. Diagram of high-temperature gas dilatometer.

However, this method does not give complete information as to the forces by which densification is attained. Furthermore, the expansion (shrinkage) of the pressed samples is not the same in all directions (in directions perpendicular and parallel to the compacting pressure). Hausner [81] has studied radial and axial shrinkage and the relationship between them. Nevertheless, the dilatometric method is widely used because it is comparatively simple, reliable, and accurate.

A high-temperature gas dilatometer has been described [12] that permits accurate measurements of shrinkage (expansion) in an atmosphere of a gas (hydrogen, argon, helium) at temperatures of 1500–1600°C. The apparatus is a differential mirror dilatometer in which the measurements are based on recording the difference in expansion (compression) of a standard and the sample. Changes in the sizes of the sample and the standard are recorded by means of a rod of quartz or other material (sapphire, corundum). The dilatometric effect is measured by recording the shift of a light spot from a plane mirror (Fig. 16).

The apparatus consists of a quartz reactor with removable brass covers. The whole apparatus is mounted on a metal table 1 with an opening for the lower water-cooled brass cover 2. Inside the reactor 3 a magnesite base 4 is placed on the lower cover. Openings are provided in the cover and the base for a thermocouple. In the top cover 9 there are two openings for steel electrical leads which are held and centered by means of Wilson seals. A vertical tubular furnace 6 is mounted on the electrical leads; vertical movement of the furnace is possible by moving the electrical leads 8, which are connected with a special device to avoid misalignment.

The heater is a tungsten cable prepared from thin (0.25–0.30 mm in diameter) wire laid in the grooves of a ceramic cylinder of beryllium oxide. The furnace is thermally insulated with four molybdenum and six nickel shields. The screens are cylinders placed one inside the other, 2–3 mm apart, with corrugated molybdenum and nickel plates placed between their walls. The length of the heating zone of the furnace is 150 mm and the diameter 30 mm.

The furnace is heated by alternating current; the voltage is measured by means of an RNO–250 variable-ratio transformer 15; the temperature is recorded by means of a thermocouple, the hot junction of which enters the furnace through an opening in the top cover of the reactor. Hydrogen purified of oxygen and water vapor enters the furnace through a connecting pipe in the head 10; the hydrogen outlet is through the connecting pipe in the lower cover.

The sample 16 is placed on a support plate of sintered aluminum oxide on top of the magnesite base. Through openings in the top cover extend two rods 7 moving freely in the vertical direction (the rods are usually of aluminum oxide). One rod rests on the sample and the other on the support plate. Mirror 14 rotating freely about its horizontal axis is placed on top of the rods. A metal bracket to which the mirror is fastened is clipped onto the end of one rod; on the end of the other rod is a thin U-shaped spring with two openings; the pivot of the mirror is attached to this spring. If the rod is deflected to one side when it is raised or lowered, the spring prevents any change in the distance between the supporting points of the mirror.

Since the coefficient of linear expansion of aluminum oxide is small and the difference in the lengths of the rods is around 15 mm, i.e., is equal to the height of the sample, with a total length of 270–280 mm, the difference in the thermal expansion of the rods can be neglected and the change in the position of the light spot can be considered to result only from changes in the linear dimensions of the sample. Rods of quartz are even better, since their coefficient of expansion is smaller still. Usually the rod is placed not directly on the sample but on a thin (0.5–0.4 mm thick) plate of aluminum oxide in order to distribute the pressure of the rod evenly over the sample.

The apparatus does not account for the effect of the pressure of the dilatometer rod on the sample, but since the weight of the sample is the same as that of the rod (not over 8 g), which is small by comparison with the capillary pressure during the sintering process, the corresponding constants will be small and can be neglected.

A narrow beam of light 12 is transmitted through a prism 13 to the sample and, reflecting from it, impinges on the scale 11 with a given curvature, which ensures that the linear displacement of the light spot on the scale is proportional to the change in the length of the sample during sintering.

The apparatus makes it possible to record changes in the linear dimensions of the sample both under isothermal conditions and during programmed heating.

With careful control of the heating it is possible to ensure a constant working temperature within limits of ±5°C.

The magnification factor of the apparatus is 150, but it can be increased considerably if necessary; the sensitivity of the apparatus is 0.01 mm. This sensitivity is completely adequate for studies of liquid-phase sintering processes.

The dilatometer described has proved to be convenient and reliable in use, since its inertia is small and it can operate at high temperatures.

In some cases [2] a dilatometer is used in which changes in the linear dimensions of the sample are measured by recording the moment when an electric circuit is closed as the result of contact of the dilatometer rod with the sample. The shrinkage or expansion of the sample is determined by means of a micrometer attached to the dilatometer rod. A diagram of this type of dilatometer is shown in Fig. 17. The sample 3 is placed vertically in a crucible 4 located within a metallic shield 5. The shield is suspended on flexible rods from the cover of the quartz reactor 6. During the measurements the screen is firmly held against the bottom of the reactor. In the holes drilled in the upper ends of the sample 3 and the reference standard 1, which are separated by a diaphragm 2, are placed the hot junctions of the thermocouples 10

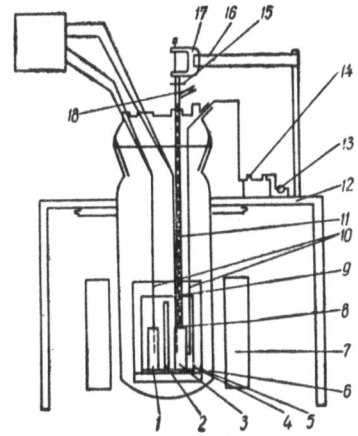

Fig. 17. Diagram of apparatus for measuring the dilatometric effect.

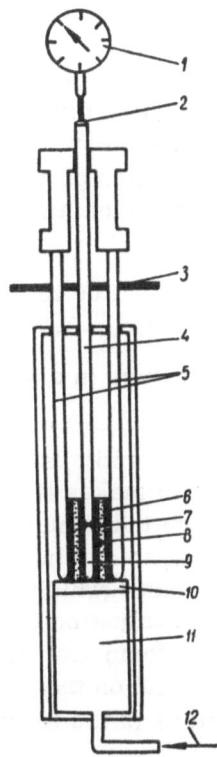

Fig. 18. Vertical section of dilatometer. 1) Indicator; 2, 10) aluminum oxide disks; 3) reflector; 4,5) silicon carbide rods; 6) graphite housing; 7) graphite crucible; 8) powder; 9) sample; 11) base (Al_2O_3); 12) gas inlet.

with protective sheaths. Through the opening in the cover passes the dilatometer rod 11, consisting of a quartz tube with a molybdenum tip 8. The tip is pressed tightly against the ground end of the tube by the spring 9, which is also part of the electric circuit. At the top of the tube there is an outlet for hydrogen, which enters the reactor through one of the openings in the cover. The rod-cover seal is a thin rubber tube. The rod is fastened to the micrometer 17, which is attached to a massive metal yoke 16. The fork 15 prevents movement of the rod during rotation of the head of the micrometer. The entire apparatus is mounted on a water-cooled metal table 12. The electric circuit for determining the moment when the rod touches the sample consists of a current source 14, an ammeter 13, and supply leads. The temperature of the furnace 7 is controlled by a temperature regulator with a thermocouple sensor.

The apparatus described is characterized by substantial inertia and therefore should be used only at temperatures up to 1000°C.

The main virtue of this apparatus is the fact that a constant pressure of the dilatometric rod on the sample is avoided, which widens the area in which it can be used. Also, the apparatus makes possible simultaneous dilatometric and differential thermal analyses.

A diagram of another widely used dilatometer [63, 103] is shown in Fig. 18.

A calibrated indicator is attached to a special fitting mounted on three recrystallized silicon carbide rods 8 mm in diameter. Changes in the length of the sample are transmitted directly to the indicator through a similar rod placed in the center and resting on the sample. The indicator makes it possible to make measurements with an accuracy of 0.01 mm. The furnace may be heated by a spiral silicon carbide heating element to a temperature of 1600°C; the temperature is measured by means of a Pt/ Pt-Rh thermocouple.

To prevent the heating of the indicator, which would induce error in the dilatometric measurements, the indicator is located at some distance from the furnace and is protected against heating by an aluminum oxide reflector. Also, a small aluminum oxide disk is placed on the end of the measuring rod, so that heat transfer through the rod is almost excluded.

Preliminary tests at 1600°C have shown that the heating of the indicator is negligible.

The effect of the weight of the rod and the indicator on the sample is not excluded in this dilatometer.

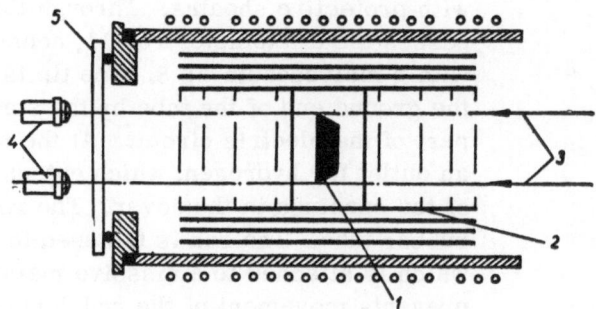

Fig. 19. Diagram of induction furnace. 1) Sample; 2) molybdenum reactor; 3) light source; 4) telescope; 5) inspection
window.

The powder to be investigated is usually pressed in a cylindrical graphite ring, which is placed between the rods. A graphite cup is placed under the lower end of the measuring rod in order to prevent the sample components from reacting with the rods.

Kingery [85] used the following method of measuring the linear dimensions of samples during sintering.

The sample, a cylinder 3 mm in diameter and about 7 mm long or a prism 7 × 3 × 4 mm in size, was placed on an aluminum oxide support plate and heated in an induction furnace with a molybdenum unit (Fig. 19). Observations were made through an inspection window. The length of the sample was determined by means of a telescope with a magnification of 9.4. The minimum measurable change in sample length was approximately 0.006 mm, which amounted to a change in length of 0.1%.

The furnace permitted heating from 700 to 1150°C in 0.5-1 min and from 1300 to 1750°C in 2 min. The experiments were made in a vacuum, in helium, or in hydrogen.

This method makes it possible to attain the required sintering temperature in a short time.

The electrical conductivity or electrical resistivity method is also frequently used for studying liquid-phase sintering. Like the dilatometric technique, this method makes it possible to investigate the sintering process in the course of time and, from the magnitudes measured (χ and R), to determine the growth of interparticle contacts during sintering, to obtain a direct record of a magnitude characterizing the surface of contact in powder compacts (which apparently cannot be achieved by any other method [28]), and a direct qualitative evaluation of the strength of the resulting contacts at different temperatures [66], to derive information on the qualitative and in some cases quantitative changes in the surface of contact and on the presence and thickness of impurity films, and to elucidate the role of the major factors in the formation of the surface of contact during sintering.

However, the electrical conductivity method makes it possible to study only the very beginning of the sintering process, when intense contact-surface formation takes place, while in the later stages of sintering the data are less precise due to small changes in the magnitude measured. Nevertheless, this method is widely used, and in combination with the dilatometric method can give the desired information on sintering processes.

The magnetometric method can be used to study the sintering process continuously in the course of time [73].

Koster and Raffelsieper [91] obtained data on the magnetic saturation of copper–nickel samples during sintering. Curves of the concentration of solid solutions obtained on the basis of magnetic measurements give a clear picture of the kinetics of homogenization. The calculated values of heat of activation for the sintering process agree with the corresponding values for the diffusion of nickel in copper.

Ananthanarayanan and Gibsch [50] found a distinct dependence of the coercive force on sintering time for mixtures of ferromagnetic and nonferromagnetic components. They show that the coercive force is a particularly sensitive parameter in the initial stages of sintering. However, the magnetometric method is suitable for investigating only those materials in which the principal or, at least, an important component is a ferromagnetic material. Furthermore, the measuring instrumentation in this technique is fairly complicated.

Schreiner and Glawitsh [116] developed a new method of investigating the sintering process based on the ionization of an inert gas surrounding a probe, induced by radioactive decomposition of an emanating radioactive substance evenly distributed in the probe (Fe–Ni and Cu–Fe systems, etc.). Schreiner and Glawitsh show that the method can be used for studying the changes in the free surface of pressed samples during isothermal sintering, when the diffusion coefficient of the emanation to the surface of the sample remains constant (disregarding any changes in the emanation concentration gradient during sintering, which places the constancy of the diffusion coefficients in doubt).

With changes in temperature it is possible to follow the ionization in relation to the diffusion coefficient in temperature ranges where no change in the free surface of the sample occurs. This latter condition reduces the value of the method, since this temperature range cannot be large. Furthermore, this method requires a complicated apparatus.

Bovarnick [54] proposed a method of investigating the sintering process based on the linear variation of the free energy of systems with the electrochemical potential in conformity with the Gibbs–Helmholtz equation. The sample is used as an electrode in an electrolytic cell and the electrochemical potential is measured. Bovarnick derived a formula relating the electrochemical potential and the densification coefficient P:

$$P = \frac{d}{d_s} \tau^{1/\prime} T^2 e^{E/RT}, \tag{47}$$

where d/d_s is the relative density, τ is time, T is temperature, and E is the electrochemical potential.

In recent years, high-temperature microscopy has been widely used [92]. This method of investigating the sintering process is very promising, since it enables the kinetics of variation in the microstructure of the sample surface at high temperatures to be followed precisely, and also the formation of solid solutions, the growth of particles of one of the components in a melt of the other or intermetallic phases, reactions leading to grain growth, and the formation of liquid phases. This method can be used to follow the process regardless of whether the final structure is heterogeneous or homogeneous.

However, high-temperature microscopy has its own special problems. Usually there is a considerable shift of the object in the field of view of the high-temperature microscope. Also characteristic is a distortion of the image induced by the convection currents between the furnace and the auxiliary lens, and there are other phenomena creating difficulties in conducting the experiments.

Most of the remaining methods of investigation, based on examination of the microstructure and measurements of the bending or tensile strength, hardness, and modulus of elasticity, do not make it possible to follow the entire course of sintering, since, in most cases, these characteristics are measured after cooling of the samples.

2. Preparation of Samples for Investigation

The compacts used for studying liquid-phase sintering processes can be prepared by different methods. The samples can be prepared by the conventional method, i.e., the powders are mixed in the desired proportions, pressed, and the compacts are sintered, or the powder of the refractory component is first pressed and then placed in contact with a given amount of the low-melting component, which at sintering temperature penetrates into the solid sample, impregnating it, thereby leading to ordinary liquid-phase sintering. This so-called impregnation method with subsequent sintering was used by the present authors to investigate some characteristics of the densification process during sintering.

Eremenko, Naidich, and Lavrinenko [17, 36] prepared samples for investigation mainly from commercially pure powders. The powders were passed through a No. 004 screen, so that the particle size was 40 μ or smaller.

To remove oxide films, the powders were first reduced in a stream of purified hydrogen for 1.5-2 h at temperatures required to remove the films. Mixtures (approximately 20 g) were prepared by thoroughly grinding the components in a porcelain mortar for 30-40 min or by mixing in a laboratory ball mill for 1-1.5 h. The powders and pressed samples were kept in a vacuum dryer or a dryer filled with hydrogen. The required porosity of the samples was obtained by calculating the batch weight and using double-ended pressing with a stop. Double-ended pressing was used to obtain a uniform density along the height of the sample. Metallic mixtures (W–Cu, Ni–Pb) were pressed with additions of ethyl alcohol, which facilitated pressing and imparted sufficient strength. Cermet mixtures with nonplastic components (TiC–Co) were mixed with a plasticizer (5% solution of rubber in benzine). The mixtures were dried in a desiccator and then granulated and pressed. The diameter of the pressed samples was 10 mm, the height 15 mm.

For impregnation and subsequent sintering the samples were placed in a crucible, to which a certain amount of the low-melting metal powder was added.

The initial and final densities were determined from the weight and size of the samples. Dilatometric measurements were made under isothermal conditions or conditions of programmed heating in an atmosphere of purified hydrogen with holding for periods established during the experiments. The heating conditions were determined by the temperature at which the liquid phase appeared and the interaction between phases.

The temperature at which noticeable shrinkage began was measured first. Ordinarily this temperature is close to the temperature at which the liquid phase appears. In the experiments the samples were heated in 15-20 min to the temperature at which shrinkage began; the temperature was then raised sharply (1-2 min) to the desired level and measurements commenced. In a number of cases the samples were placed in a preheated furnace. In this case the results differed only little, since the heating was rapid (1-1.5 min). In all cases at least two parallel experiments were run and kinetic curves were plotted from the results.

3. Method of Determining Wettability
in Systems Undergoing Sintering

To clarify the role of wetting in densification processes during liquid-phase sintering, the wettability of solid surfaces in liquid metals and alloys was investigated at different temperatures [18, 36]. Since wetting is affected by various impurities and additions, the wettability experiments were conducted in the same apparatus used for the sintering experiments in order to obtain experimental conditions as close as possible to those used in the shrinkage experiments. The temperature and the medium matched those used in the sintering experiments. The angle of contact was measured on a solidified drop in an optical apparatus permitting measurements with an accuracy of ±1° [15].

The experiments were conducted in the following manner. On the flat surface of a support plate from the test material was placed a pressed cylindrical metal sample (of the same purity as in the sintering experiments). The sample was then placed in the furnace and held at a given temperature. The angle of contact at each temperature was measured after holding for 30 min.

Literature data [53] and our observations indicate that this holding time is sufficient to obtain consistent values of the angle of contact. Since the drops were not always symmetrical, the angle of contact θ was measured in several directions and the average value was determined. The average value from two to three parallel experiments was used as the true angle of contact.

The values of θ determined on the solidified drop may differ somewhat from those obtained directly in the heating process due to shrinkage of the drop during cooling. It was shown in [21] that such a difference in size is in fact observed, but does not exceed 3-10%. Therefore, angles of contact determined from the shape of a solidified drop are quite adequate for qualitative analysis of the effect of wetting on sintering kinetics in the presence of a liquid phase.

CHAPTER 4

DENSIFICATION PROCESSES DURING SINTERING IN SYSTEMS WITH COMPONENTS INSOLUBLE IN THE LIQUID AND SOLID STATES

1. Tungsten—Copper System

According to literature data, copper and tungsten are practically insoluble in each other either in the liquid or solid states [48]. However, solid tungsten is readily wetted by liquid copper.

Some results from a study of tungsten—copper alloys were given by Cannon and Lenel [59], who studied the effect of alloy composition (5 and 10 wt.% Cu), the density, and the gaseous medium on densification during sintering (Fig. 20). They found that in samples of 90 wt.% W and 10 wt.% Cu compacted under a pressure of 700 daN/cm^2, with an initial density of 6.6 g/cm^3 (curve 3, Fig. 20) the density increased to 11.6 g/cm^3 after sintering in hydrogen for 30 min at 1310°C. Samples compacted under a pressure of 1400 daN/cm^2, with an initial density of 7.4 g/cm^3 (curve 4, Fig. 20), had a density of 8.35 g/cm^3 after sintering under the same conditions. Price and his co-workers [105] found no densification in samples of 93 wt.% W + 7 wt.% Cu with an initial density of 10.5 g/cm^3. Detectable densification was observed only in samples in which the initial tungsten particle size was 3 μ. No densification was observed in samples with larger particles under the same conditions.

A detailed study of densification in the W—Cu system in relation to various factors was conducted by Naidich, Lavrinenko, and Eremenko [36], in which the tungsten particle size was 1-5 μ and the copper particle size was 40 μ or less. In some cases the total shrinkage curve was plotted and in others, where only the maximum shrinkage was of interest, the size of the samples was measured only before and after sintering.

The choice of compositions was dictated by the necessity of studying the kinetics of densification at widely varying ratios of the components for the purpose of determining the effect of the low-melting component concentration on the rate and magnitude of densification and of establishing the most suitable compositions for maximum densification. Since a low initial porosity does not make it possible to follow the process because of various secondary phenomena connected with the high compacting pressure and frequently has an adverse effect on densification, the samples prepared for the investigation had a fairly high initial porosity.

The kinetics of the densification process were investigated on samples with initial porosities of 40 and 50% and containing 20, 30, 40, and 50 vol.% copper, since at higher

TABLE 1. Slope k of Densification Curves for Samples of W—Cu System

Alloy	Temperature, °C	
	1150	1350
Initial porosity 50%		
W—50 vol.% Cu	1.51	1.9
W—30 vol.% Cu	—	1.07
Initial porosity 40%		
W—40 vol.% Cu	1.01	1.38
W—30 vol.% Cu	0.83	—

30

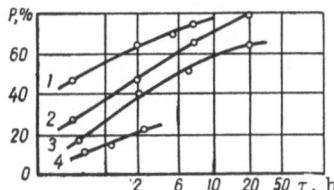

Fig. 20. Variation of densification coefficient of tungsten–copper samples with sintering time in different media. 1) 90 wt.% W + 10 wt.% Cu, d = 6.6 g/cm^3 (vacuum); 2) 85 wt.% W + 15 wt.% Cu, d = 6.6 g/cm^3 (hydrogen); 3) 90 wt.% W + 10 wt.% Cu, d = 6.6 g/cm^3 (hydrogen); 4) 90 wt.% W + 10 wt.% Cu, d = 7.4 g/cm^3 (vacuum).

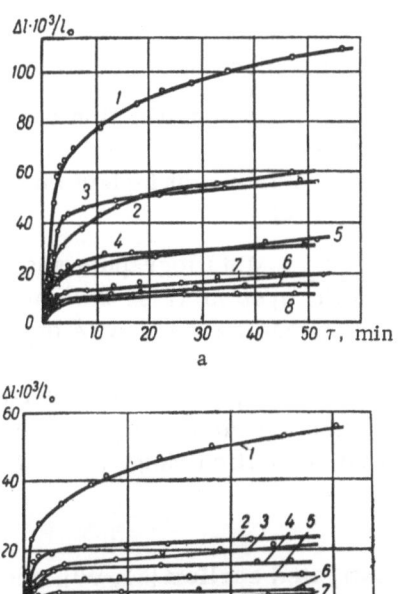

Fig. 21. Dilatometric densification curves of W–Cu samples with an initial porosity of 50% (a) and 40% (b). 1) 50 vol.% Cu, 1350°C; 2) 50 vol.% Cu, 1150°C; 3) 40 vol.% Cu, 1350°C; 4) 30 vol.%, 1350°C; 5) 40 vol.% Cu, 1150°C; 6) 30 vol.% Cu, 1150°C; 7) 20 vol.% Cu, 1350°C; 8) 20 vol.% Cu, 1150°C.

copper concentrations (60%) samples are deformed during heating. The measurements were made in a hydrogen atmosphere at temperatures of 1150, 1350, and in some experiments at 1250, 1450, and 1550°C. The error in measuring the shrinkage was ±2%.

In order to determine the dependence of the densification process on time the curves obtained (Fig. 21) were plotted as straight lines in logarithmic coordinates and their slope was determined (Table 1).

As a result, it was established that in the initial stage of sintering the shrinkage is satisfactorily described by the relationship $\Delta l / l_0 \sim \tau^k$, where k = 1-2, and then the shrinkage decreases sharply and ceases completely.

Effect of Temperature. As shown by the investigations, with different porosities the shrinkage gradually increases with increasing copper concentration and rising sintering temperature. Only at a temperature of 1350°C for the W–50% Cu alloy does shrinkage increase stepwise irrespective of the initial porosity of samples. To study the shrinkage at a temperature above 1350°C (i.e., a temperature at which complete wetting of tungsten with copper is observed) and to eliminate the effect of vaporization of copper, the holding time was 15 min (Fig. 22).

The results indicate that the shrinkage of samples varies most in the temperature range 1250–1350°C.

Effect of Quantity of Liquid. Investigations in this area were conducted by two methods [36]: by measuring the shrinkage during and after sintering, and also by measuring the shrinkage during the impregnation of a porous skeleton with liquid metal and subsequent sintering.

In the former case samples were prepared with a constant concentration of the refractory component and different concentrations of the low-melting component and were compacted under different pressures, so that the volumes of the low-melting component and the pores were constant. The samples contained 32 vol.% W and 12, 16, 24, 32, 36, 40, 44, 48 vol.% Cu.

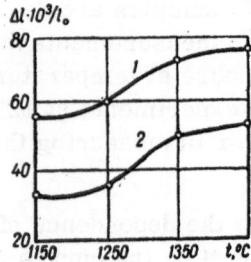

Fig. 22. Variation of shrinkage of W–Cu samples with a porosity of 50% (1) and 40% (2) in relation to the concentration of liquid phase. 1) 40 vol.% Cu; 2) 50 vol.%.

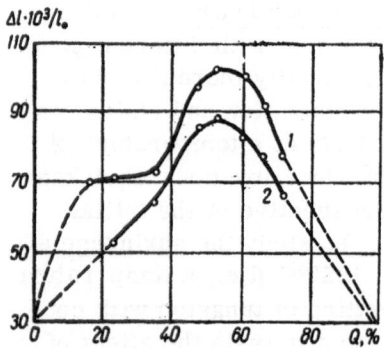

Fig. 23. Variation of shrinkage of W–Cu samples with the degree of filling of pores with liquid phase. 1) During impregnation; 2) during sintering.

As a consequence of the fact that with increasing amounts of the low-melting component it was necessary to apply higher pressures, interlocking of the solid particles could not be ruled out, which could have affected shrinkage. Therefore, the method of measuring shrinkage during the impregnation of samples and subsequent sintering was adopted. Samples of pure tungsten were compacted to a given porosity (68 vol.%) and then placed in a crucible to which different amounts of the low-melting component were added, thereby controlling the degree of filling of the pores by the liquid.

The samples were impregnated and sintered in hydrogen at 1300°C with holding for 30 min. During impregnation the tungsten samples absorbed all the copper.

The initial and final sizes of the samples were determined with a micrometer.

According to the curves shown in Fig. 23, with increasing amount of the liquid phase the shrinkage at first increases, reaches a peak, and then decreases. The curve for impregnation with subsequent sintering lies somewhat above the curve for sintering without impregnation. The occurrence of a peak on a curve of shrinkage plotted against the extent of pore filling with the liquid phase will be considered below.

Effect of Porosity. The densification coefficients of W–Cu alloys are given in Table 2. Higher values of densification coefficient were obtained at higher initial porosities for all compositions [59].

The Microstructure. Metallographic analysis of sintered samples of the W–Cu system (Fig. 24) showed that a rise in the temperature promotes the flow of the liquid phase around the tungsten grains. This indicates a change in the wetting conditions and in the energy at the interphase boundaries. No substantial grain growth of tungsten was noted in the temperature and composition ranges investigated.

2. Tungsten–Silver System

Several alloys of the tungsten—silver system were investigated earlier [51, 90], and it was shown that samples containing 60–70 vol.% W and 30–40 vol.% Ag sintered 3–4 h at 1000°C are characterized by high hardness, strength, etc.

According to Langraf [96], nonporous sintered tungsten–silver alloys can be obtained from completely oxide-free powders under pressure in a hydrogen atmosphere. Such alloys exhibit good ductility properties. Silver and tungsten are mutually insoluble both in the liquid and solid phases [48]. Therefore, alloys of these metals are usually prepared by the powder metallurgy method — sintering in the presence of a liquid phase or impregnation of a porous skeleton with liquid metal.

A detailed study of densification in the liquid-phase sintering of the W–Ag system was conducted by Naidich, Lavrinenko, and Eremenko [36]. Dilatometric densification curves were obtained at temperatures of 1000, 1100, and 1200°C, and in some cases at 1300°C, for samples

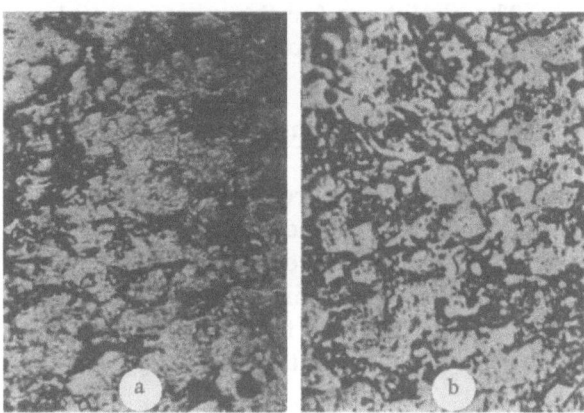

Fig. 24. Microstructure of W + 50 vol.% Cu. Sintering temperatures 1150°C (a) and 1350°C (b). × 450.

TABLE 2. Densification Coefficients for W–Cu Samples

Alloy	Temperature, °C	
	1150	1350
Initial porosity 50%		
W — 20 vol. % Cu	3.5	4.5
W — 30 vol. % Cu	5.1	9.8
W — 40 vol. % Cu	12.5	16.5
W — 50 vol. % Cu	17.8	37.4
Initial porosity 40%		
W — 20 vol. % Cu	0.8	1.0
W — 30 vol. % Cu	1.8	4.1
W — 40 vol. % Cu	6.2	8.3
W — 50 vol. % Cu	14.7	27.0

with different silver concentrations (20, 35, 50 vol.%).

Sintering was conducted either at constant temperature or with stepped heating. In the latter case the heating conditions were as follows. The samples were heated in 15-20 min to temperatures of 870–900°C (at which no shrinkage is as yet observed); the temperature was then rapidly (0.5–1 min) raised to 1000°C, the sample being held at this temperature until shrinkage ceased, after which the temperature was again increased to 1100°C, etc. Such a procedure clearly revealed the effect of temperature on shrinkage (Fig. 25).

To establish the variation of shrinkage with time, the dilatometric curves were plotted as straight lines in logarithmic coordinates and their slopes were determined.

For W–Ag samples with an initial porosity of 50% and a sintering temperature of 1000°C the slopes of the lines k were as follows:

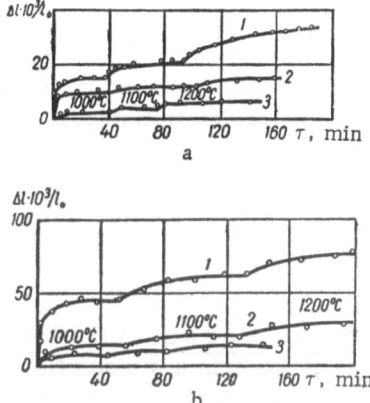

Fig. 25. Variation of shrinkage of W–Ag samples with sintering time at an initial porosity of 40% (a) and 50% (b). 1) W + 50 vol.% Ag; 2) W + 35 vol.% Ag; 3) W + 20 vol.% Ag.

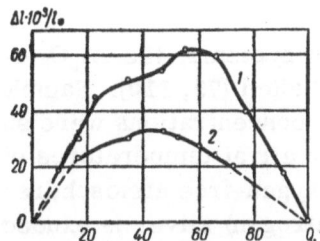

Fig. 26. Variation of shrinkage in the W–Ag system with the degree of filling of pores with liquid phase (temperature 1200°C). 1) Impregnation; 2) sintering.

Alloy	k
W + 50 vol. % Ag	1.6
W + 35 vol. % Ag	1.27
W + 20 vol. % Ag	1.13

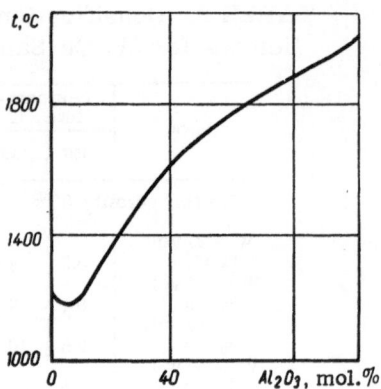

Fig. 27. Melting diagram of the $Al_2O_3-Cu_2O$ system.

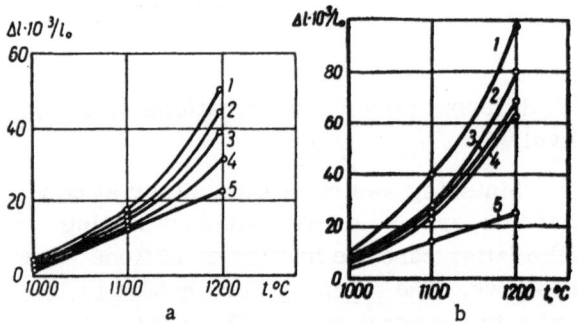

Fig. 28. Variation of shrinkage of $Al_2O_3-Cu_2O$ samples with sintering temperature in sintering for 1 h (a) and 5 h (b). 1) 50 vol.% (Ag + 10 wt.% CuO); 2) 50 vol.% (Ag + 5 wt.% CuO); 3) 30 vol.% (Ag + 10 wt.% CuO); 4) 30 vol.% (Ag + 5 wt.% CuO); 5) 30 vol.% Ag.

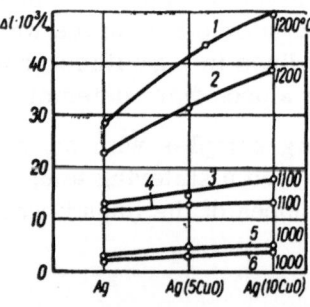

Fig. 29. Variation of shrinkage of $Al_2O_3-Ag-CuO$ samples with the concentration of CuO and Ag.

The shrinkage for this system is also described by the relationship $\Delta l/l_0 \sim \tau^k$, where k = 1-2.

Effect of Temperature. An investigation showed that the shrinkage increases with rise in temperature regardless of the silver concentration or the porosity. It should be noted that, at a concentration of 35-50 vol.% silver, the jumplike increase of shrinkage is particularly noticeable for alloys of high initial porosity (50%) (cf. the W-Cu system).

Effect of the Quantity of Liquid. Experiments were conducted to determine the shrinkage after ordinary sintering and also after impregnation of porous tungsten with liquid silver and subsequent sintering at 1200°C with holding for 30 min. As can be seen from Fig. 26, a maximum occurs in the region of 50% filling of the pores with the liquid phase. Similar results were obtained also for the W-Cu system. The densification coefficient also varies with the concentration of liquid silver in the samples (temperature 1200°C):

Ag, vol. %	11	23	35	47	52	66	72	88
P, %	4.7	11.7	19.8	29.8	39.2	46	44	29.5

Effect of Porosity. The initial porosity of samples has a substantial effect on their shrinkage. Increasing porosity results in an increase of shrinkage (see Fig. 24).

3. Other Systems

Other metallic systems of mutually insoluble components have not been investigated systematically.

Of cermet systems, the Al_2O_3-Ag system has been studied [79, 110]. Samples with different silver concentrations were successfully sintered in air at temperatures of 1000-1250°C; in an oxygen-free atmosphere (argon, hydrogen, or nitrogen) silver is exuded by the samples. As is well known, oxygen dissolves in liquid silver in substantial quantities and produces a marked increase in the strength of the metal—oxide bond [15, 110].

Kingery and co-workers [87] have investigated the Al_2O_3–glass system (72 vol.% SiO_2, 15 vol.% Na_2O, 9 vol.% CaO, 3 vol.% MgO, 1 vol.% Al_2O_3). Samples were prepared from powdered glass with a particle size of about 5 μ. The samples were sintered in air with rapid heating to the sintering temperature.

Under the experimental conditions employed, the components were insoluble and no grain growth of the solid phase or marked densification was observed. Most of the densification occurred in the first hour of sintering, although some densification was observed even after 160 h.

Examination of the microstructure showed that the final sample density depends on the homogeneity of the mixture and the heating conditions. The initial size of the aluminum oxide particles remains unchanged, and the particle edges and points remain sharp, which indicates low solubility.

We investigated sintering in the Al_2O_3–Ag–CuO system without a protective atmosphere (in air).

It is well known that aluminum oxide does not react with silver and is not dissolved in it. Cuprous oxide and aluminum oxide [121] form a liquid phase at a temperature of about 1180°C (Fig. 27). At high temperatures cupric oxide transforms into cuprous oxide.

A horizontal tubular furnace was used in the investigation; shrinkage was determined from micrometric measurements of the initial and final sample dimensions. The initial porosity of all samples was 50 ± 1.5%.

Samples were prepared by thoroughly mixing the silver and cupric oxide powders in the given ratios, and then aluminum oxide powder was added in the desired quantity and the powders were again mixed for 40–45 min. Benzyl alcohol was used as a plasticizer in compacting to obtain compacts of higher strength. Samples of the following compositions were prepared: Al_2O_3 + 30 vol.% Ag, Al_2O_3 + 50 vol.% Ag, Al_2O_3 + 50 vol.% (Ag + 5 vol.% CuO), Al_2O_3 + 50 vol.% (Ag + 10 vol.% CuO), Al_2O_3 + 30 vol.% (Ag + 5 vol.% CuO), and Al_2O_3 + 30 vol.% (Ag + 10% CuO).

Sintering was conducted at 1000, 1100, and 1200°C for 1 and 5 h.

It can be seen from Fig. 28 that with rise in temperature the shrinkage increases, particularly at 1200°C. The increase of shrinkage with rising temperature in this case may be due to two factors: 1) a change in the angle of contact, leading to an increase of the capillary pressure between particles; 2) the formation of additional liquid at 1200°C as a result of reaction between the Al_2O_3 and CuO. This latter hypothesis is confirmed by the fact that, at a constant aluminum oxide concentration (vol.%), the shrinkage is greater in samples with a higher CuO concentration. It should be noted that with increasing temperature the shrinkage increases also in Al_2O_3 + 30 vol.% Ag samples. This is particularly pronounced at a temperature of 1200°C.

The investigations have shown that the sintering at 1000 and 1100°C for 1 h of samples containing CuO leads to a weight change of no more than 1–1.5% of the weight of the silver in the sample.

In Al_2O_3 + 30 vol.% Ag samples sintered at 1000 and 1100°C, the change in weight is also small, while in alloys containing 50 vol.% Ag sweating out of the metal is observed over the entire surface of the sample, the concentration of silver being thereby reduced to 31–31.5 vol.%.

Samples of Al_2O_3 and pure silver are brittle and easily disintegrate, while samples of the Al_2O_3–Ag–CuO system have considerable strength. Thus, it can be concluded that CuO promotes the formation of stronger bonds.

Effect of Quantity of Liquid.   As can be seen from Fig. 29, the shrinkage of alloys as a function of their CuO concentration varies only negligibly at 1000 and 1100°C, while
at 1200°C it increases sharply as a result of reaction between Al_2O_3 and CuO, which is accompanied by the formation of a liquid phase.

CHAPTER 5

DENSIFICATION PROCESSES IN SYSTEMS WITH
LIMITED SOLUBILITY OF THE COMPONENTS

1. Nickel—Titanium Carbide System

From the phase diagram of the pseudobinary Ni—TiC system [20] (Fig. 30) it can be seen that the eutectic temperature is 1280°C at 9.3% titanium carbide. The solubility of titanium carbide in nickel in the solid state at this temperature is 6.2% and gradually decreases to 2% at 700°C. Sintering in this system usually occurs in the presence of the liquid phase.

The sintering process in the Ni—TiC system was investigated in detail by Niki et al. [101, 102]. They studied the effect of the amount of metallic binder on shrinkage and also the effect of temperature, gaseous medium, and amount of metal on the size and shape of the grains. The results, however, cannot be compared with the well-known theories, since the experimental conditions were not comparable. The samples were heated to 1500°C at the rate of 250°C/h, held at this temperature about 1 h, and cooled slowly. Thus, the time for attaining the temperature at which the liquid phase appeared was too long, as was also the time for reaching 1500°C. Therefore, it is impossible to follow the rate of this densification process.

Some data on the sintering of cermets based on titanium carbide with a nickel binder can be found in [26].

The densification kinetics of samples of this system with 5, 17, 28, and 35 vol.% Ni at an initial porosity of 30 and 40% was investigated in [17]. Measurements were made at temperatures of 1300, 1350, 1420, and 1500°C in a hydrogen atmosphere with holding for 1 h. The samples were heated to the sintering temperature as follows: to 1000°C in 20 min, followed by a rapid increase to the given temperature; temperatures of 1300–1420°C were reached in 4–6 min, and 1500°C in 10–12 min.

Figure 31 shows some of the kinetic densification curves. Shrinkage is slight up to 1000–1100°C (depending on the composition) and then, at the moment when the liquid phase appears (1280°C), it increases sharply; it is a maximum at the beginning of sintering. The shrinkage increases with the temperature, porosity, and amount of liquid phase in the sample. The curves obtained were plotted as straight lines in logarithmic coordinates and the slopes of the lines were determined (Table 3).

One observes two straight-line sections on the plots of log $(\Delta l/l_0)$ vs. log τ. In the first, $\Delta l/l_0 \sim \tau^k$

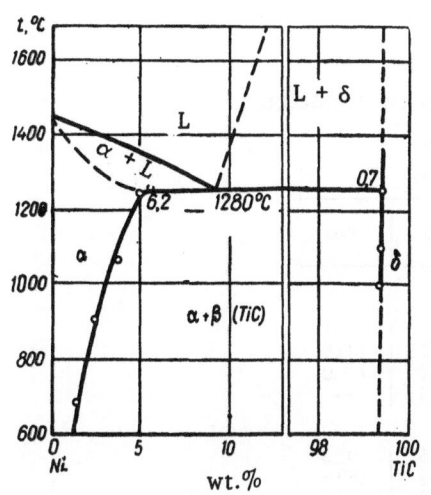

Fig. 30. Phase diagram of Ni — TiC system.

37

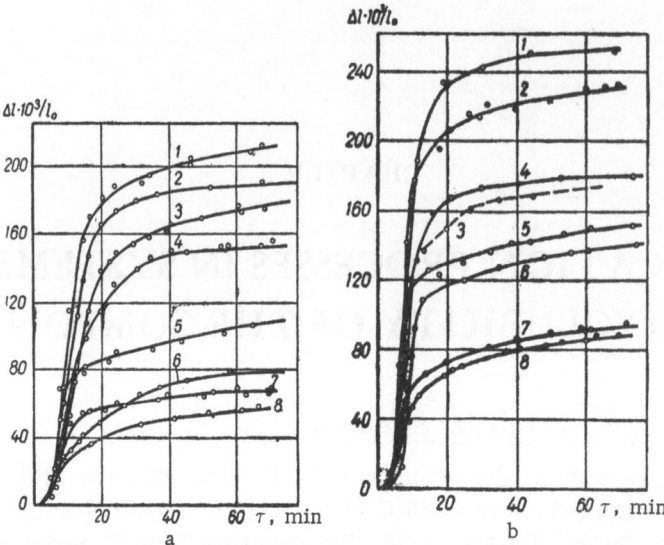

Fig. 31. Variation of shrinkage of Ni−TiC samples with sintering time at 1350°C (a) and 1500°C (b) and different Ni concentrations (vol.%): 1) 35; 2) 28; 3) 35; 4) 28; 5) 17; 6) 17; 7) 5; 8) 5. The initial porosity was 40% (1,2,5,7) and 30% (3,4,6,8).

TABLE 3. Slope k of Straight Lines in Coordinates of log $(\Delta l/l_0)$ vs log τ for Samples of Ni−TiC System

Alloy	Temperature, °C			
	1350		1500	
	k_1	k_2	k_1	k_2
TiC + 35 vol.% Ni	2.3	0.14	3.1	0.14
TiC + 28 vol.% Ni	2.15	0.28	2.5	0.16
TiC + 17 vol.% Ni	1.45	0.32	2.8	0.2
TiC + 5 vol.% Ni	—	—	2.4	0.3

TABLE 4. Shrinkage of Ni−TiC Samples in Relation to Sintering Temperature, %

Alloy	Temperature, °C			
	1300	1350	1420	1500
Initial porosity 30%				
TiC — 5 vol.% Ni	38.5	54.5	75.0	86.1
TiC — 17 vol.% Ni	74.3	77.0	106.2	138.5
TiC — 28 vol.% Ni	84.5	150.5	157.4	179.0
TiC — 32 vol.% Ni	90.8	173.5	182.0	—
Initial porosity 40%				
TiC — 5 vol.% Ni	51.0	67.8	84.5	92.3
TiC — 17 vol.% Ni	82.3	102.5	142.3	150.5
TiC — 28 vol.% Ni	132.1	187.0	188.2	230.4
TiC — 32 vol.% Ni	193.8	207.3	217.0	252.1

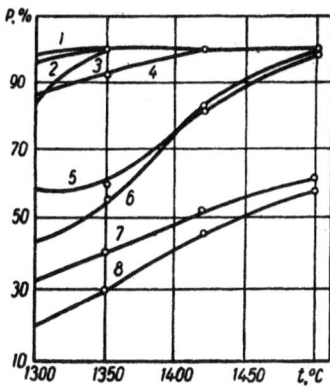

Fig. 32. Variation of density with sintering temperature, composition, and porosity of Ni–TiC samples. The Ni concentration (vol.%) was: 1,2) 35; 3,4) 28; 5,6) 17; 7,8) 5. The initial porosity was 40% (1,3,6,8) and 30% (2,4,5,7).

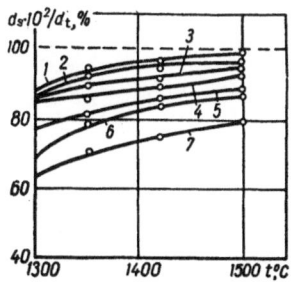

Fig. 33. Variation of densification coefficient of Ni–TiC alloys with composition, temperature, and porosity. Ni concentration (vol.%): 1), 35; 2,4) 28; 3) 17; 5,7) 5; 6) 17. The initial porosity was 30% (1,2, 3,5) and 40% (4,6,7).

(where k = 1–3) and in the second, $\Delta l / l_0 \sim \tau^k$ (where k = 0.1–0.35).

Effect of Temperature. Data concerning the effect of sintering temperature on the shrinkage of samples of the TiC—Ni system are given in Table 4.

The linear shrinkage of the TiC + 35 vol.% Ni alloy (porosity 30%) at 1500°C could not be measured precisely due to a marked deformation of the samples during sintering.

As the investigation showed, an increase of temperature promotes densification of samples of all compositions and different porosities.

Effect of Amount of Liquid. With increasing amount of the metal binder the shrinkage increases.

The variation of the density of samples with sintering temperature and the densification coefficients are shown in Figs. 32 and 33. It can be seen from the figures that, with small amounts of liquid phase, an increase of temperature does not lead to substantial densification. When the amount of liquid phase is adequate (17 vol.% or more) the residual porosity is no more than 10% after holding for 1 h at 1350°C. Raising the temperature or increasing the sintering time to 1.5–2 h results in almost pore-free samples.

In spite of the fact that shrinkage increases somewhat with increasing initial porosity, it has almost no effect on the total extent of densification, while at small amounts of liquid the densification coefficient is actually higher for alloys with lower initial porosities.

The Microstructure. Comparison of the micrographs of TiC—Ni alloys with different amounts of nickel sintered for 1 h at 1350 and 1500°C (Fig. 34) indicates that the size and shape of the titanium carbide grains depend to a considerable extent on the amount of liquid phase present during sintering, while an increase of the sintering temperature (and consequently an increase in the amount of liquid phase) promotes grain growth. At high nickel concentrations (28 and 35 vol.%) the titanium carbide grains are predominantly rounded and almost all the grains are separated from each other by the cementing metal. At low concentrations of the metal binder (5 and 17 vol.%) a large number of grains have irregular angular shapes and the carbide grains are joined.

Also, at high concentrations of liquid metallic phase the carbide grains are more uniform in shape and size than at low concentrations of liquid phase (for example, with 17 vol.% Ni).

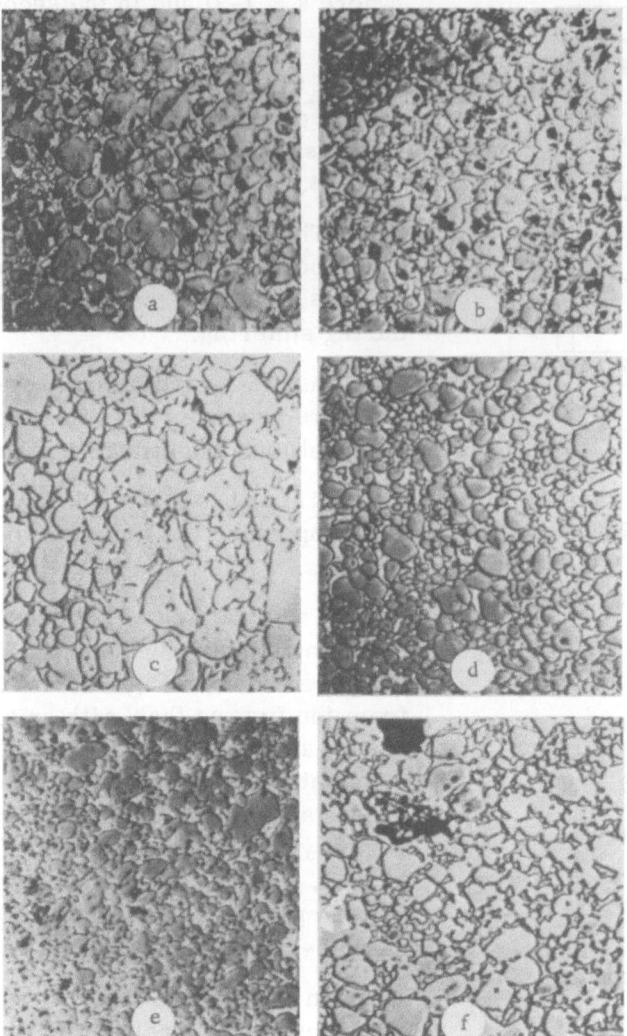

Fig. 34. Microstructures of Ni−TiC alloys sintered at
1500°C (a,b,c) and 1350°C (d,e,f). a) TiC + 35 vol.% Ni;
b) TiC + 28 vol.% Ni; c) TiC + 17 vol.% Ni; d) TiC + 35
vol.% Ni; e) TiC + 28 vol.% Ni; f) TiC + 17 vol.% Ni.
× 450.

2. Cobalt−Titanium Carbide System

Alloys of cobalt with titanium carbide undergo crystallization of the eutectic type [13, 27]
(6% Co at 1360°C). The phase diagram of this system is shown in Fig. 35.

A detailed study of the sintering process in the Co−TiC system was made in [76, 84]
and of the properties of sintered TiC−20 wt.% Co samples − shrinkage, density, hardness, and
some mechanical characteristics − in [84]. The structure and sintering mechanism of samples
with 10 and 24% Co were investigated in [76].

We investigated the densification process during the sintering of samples containing 28
vol.% Co with a porosity of 40% at temperatures of 1330, 1380, 1430, 1490, and 1530°C. Since,
judging from the phase diagram, the sintering process in the Co−TiC system is similar to
the process in the Ni−TiC system, we did not make a detailed study of the former system

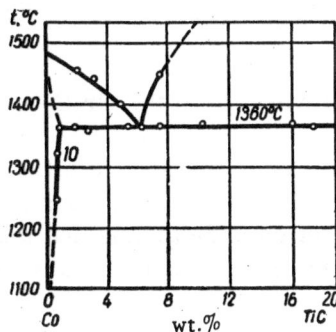

Fig. 35. Phase diagram of Co–TiC system.

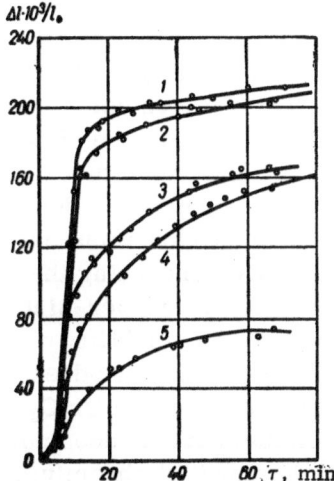

Fig. 36. Variation of shrinkage of Co–TiC samples with sintering time at different temperatures: 1) 1530°C; 2) 1490°C; 3) 1430°C; 4) 1380°C; 5) 1330°C.

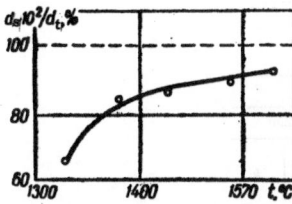

Fig. 37. Variation of density of TiC + 28 vol.% Co samples sintered at different temperatures for 1 h.

TABLE 5. Slopes of Straight Lines in Coordinates $\log (\Delta l/l_0)$ vs $\log \tau$ for Co–TiC Samples

Alloy	Temperature, °C							
	1380		1430		1490		1530	
	k_1	k_2	k_1	k_2	k_1	k_2	k_1	k_2
TiC–28 vol.% Co	1.13	0.36	1.2	0.23	1.45	0.17	1.6	0.14

but limited the investigation to samples of one composition at different temperatures. To check the possibility and conditions of obtaining pore-free samples of this system we selected a composition with a substantial amount of the metal. The experiments were made in a hydrogen atmosphere with holding for 1 h.

Effect of Temperature. The variation of the shrinkage of Co–TiC samples with sintering time at different temperatures is shown in Fig. 36. Shrinkage begins at 1200°C and increases gradually with rise in temperature. At 1360°C there is a sharp increase of shrinkage due to the appearance of the liquid phase. The shrinkage then slows down and continues for a long time.

The shrinkage curves were plotted as straight lines in logarithmic coordinates and their slopes were determined (Table 5). It can be seen from the table that there are two straight-line sections in the coordinates $\log (\Delta l/l_0)$ vs $\log \tau$ which are described by the expressions obtained for the Ni–TiC system.

The shrinkage curve for the temperature of 1330°C differs greatly from the other curves (see Fig. 36), since at 1330°C the liquid phase is still not formed, i.e., solid-phase sintering occurs.

An increase of the temperature above the eutectic sharply changes the rate and magnitude of the shrinkage.

The variation of d_s/d_t with sintering temperature, shown in Fig. 37, indicates that in solid-phase sintering (below 1380°C) samples of the given composition attain a density of no more than 65% of the theoretical density. The appearance of the liquid phase increases the densification rate and the final density. Thus, at 1380°C, the samples attain densities up to 85% and at 1530°C up to 92–93%. An increase of sintering time from 1 to 2 h results in almost pore-free samples.

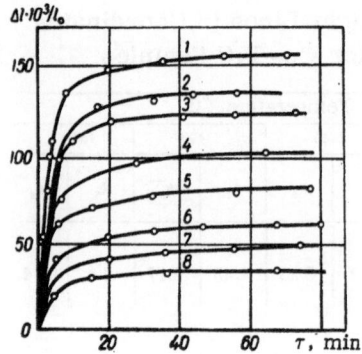

Fig. 38. Variation of shrinkage of Cr_3C_2 + 50 vol.% (Cu–Ni) samples with sintering time and temperature. 1,2) Cu + 20 vol.% Ni; 3,4,7) Cu + 10 vol.% Ni; 5,6,8) Cu. Sintering temperature: 1350°C (1,3,5); 1250°C (2,4,6); 1150°C (7,8).

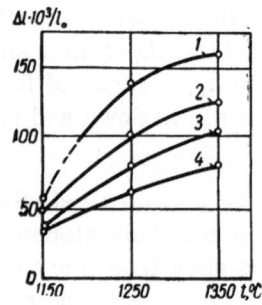

Fig. 39. Variation of shrinkage of Cr_3C_2–Cu–Ni samples with sintering temperature. 1) Cu + 20 vol.% Ni; 2) Cu + 10 vol.% Ni; 3) Cu + 5 vol.% Ni; 4) Cu.

3. Chromium Carbide — Copper — Nickel System

Grigor'eva et al. [9] investigated the Cr_3C_2–Ni system and found that in alloys with a low nickel concentration the metallic phase does not separate the carbide grains from each other. The alloys consist of a rigid continuous skeleton of carbide grains and the metallic phase is distributed in this skeleton as separate islands. A carbide skeleton was found to form in alloys with 5, 10, 15, and 20 vol.% Ni.

In alloys with 30% Ni a considerable number of grains on the face of the microsection proved to be completely surrounded by metallic phase, while in alloys with 40% Ni the carbide grains were separate and disseminated in the metallic matrix.

Eremenko et al. [18] investigated densification during the sintering of Cr_3C_2–Cu–Ni samples. The presence of copper in this system lowers the sintering temperature. This is probably a system with a limited solubility of the components. Pure copper evidently dissolves only negligible amounts of chromium carbide. There are no data in the literature concerning the solubility of chromium carbide (Cr_3C_2) in copper-nickel alloys, but it can be assumed that the solubility of chromium carbide increases when nickel is added to the alloys. The eutectic temperature of the Cr_3C_2–Ni system is 1175°C [83]. Chromium carbide does not dissolve nickel, but a substantial amount of chromium carbide is dissolved in nickel in the solid state [22]. About 12 wt.% chromium carbide dissolves in nickel at 1250°C [70].

As yet there are no data in the literature concerning sintering processes in the Cr_3C_2–Cu–Ni system.

TABLE 6. Slopes of Straight Lines in Coordinates of $\log (\Delta l / l_0)$ vs $\log \tau$ for Samples of the Cr_3C_2–Cu–Ni System

Alloy	Temperature °C					
	1150		1250		1350	
	k_1	k_2	k_1	k_2	k_1	k_2
Cr_3C_2 — 50 vol. % (Cu — 5 vol. % Ni)	1.67	0.13	1.7	1.13	2.6	0.12
Cr_3C_2 — 50 vol.% (Cu — 10 vol. % Ni)	1.75	0.16	2.05	0.13	3.8	0.11
Cr_3C_2 — 50 vol.% (Cu — 20 vol. % Ni)	—	—	3.0	0.14	4.2	0.11
Cr_3C_2 — 50 vol. % (Cu — 30 vol. % Ni)	—	—	3.5	0.11	4.2	0.07

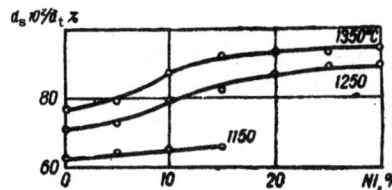

Fig. 40. Variation of the density of sintered Cr_3C_2–Cu–Ni samples with the nickel concentration.

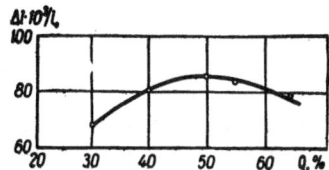

Fig. 41. Shrinkage of Cr_3C_2–Cu–Ni samples in relation to degree of filling of pores with liquid at a sintering temperature of 1250°C. Initial porosity 50%.

Densification kinetics for this system were studied at temperatures of 1150, 1250, and 1350°C. The samples were chromium carbide with 50 vol.% of copper–nickel alloys in which the nickel concentration was 0–30 vol.%. The initial porosity was 40 ± 1.5%.

The samples were prepared by pressing mixtures of the powders in the desired proportions and sintering was conducted in hydrogen for 1 h at the different temperatures.

The dilatometric data obtained are shown in Fig. 38.

At 1350°C the samples with 25–30 vol.% Ni reach a high density, the residual porosity amounting to 4–5%.

The curves shown in Fig. 38 were plotted in logarithmic coordinates and the slopes of the resulting straight lines were determined. The values of the slopes k_1 and k_2 (Table 6) indicate that the first stage of sintering is described by the relationship $\Delta l / l_0 \sim \tau^{1-4}$ and the second by $\Delta l / l_0 \sim \tau^{0.1-0.2}$.

Effect of Temperature. Figure 39 shows the linear shrinkage of Cr_3C_2–Cu–Ni samples with an initial porosity of 40% which were sintered for 1 h at different temperatures.

In all cases the shrinkage increases with rise in sintering temperature. An increase of the nickel concentration in the copper–nickel alloy also substantially increases the shrinkage.

At nickel contents up to 20% (Fig. 40), the porosity of the alloy at temperatures of 1250 and 1350°C decreases sharply and then changes by only 2–3%. It is interesting that nickel has a marked effect on shrinkage at temperatures of 1250 and 1350°C but almost none at 1150°C. This is explained as follows. An increase of the nickel concentration in the liquid phase increases the wetting of the solid by the liquid, the solubility of the solid phase in the liquid, and apparently the solubility of nickel in solid chromium carbide. The first two factors accelerate densification. The solution of nickel in chromium carbide reduces the amount of liquid phase, thus reducing the amount of nickel taking part in the sintering process. At temperatures above the eutectic all three factors increase the shrinkage.

It was of interest to determine how the shrinkage during sintering varies with the amount of liquid filling the pores in such a system. The shrinkage was measured after impregnation with subsequent sintering. During sintering, the amount of liquid in the samples varied from 30 to 65%, the nickel concentration remaining constant at 15%. The impregnating alloy was placed on the upper ends of the samples. Impregnation and sintering were conducted in hydrogen at 1250°C for 1 h.

The chromium carbide samples completely absorbed all the alloy.

As can be seen from Fig. 41, the densification was highest at around 50 vol.% of the liquid phase.

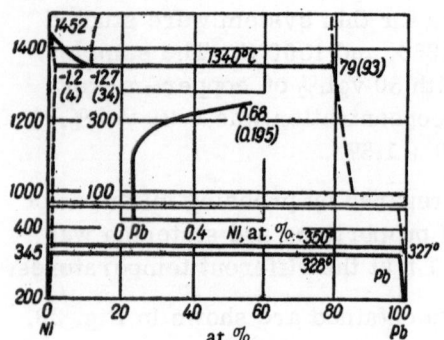

Fig. 42. Phase diagram of Ni–Pb
system.

4. Nickel—Lead System

Lead and nickel form a system with layering in the liquid state [48]. The solubility of nickel in liquid lead is about 7 wt.% (20 at.%) at 1340°C and decreases with lowering temperature. At the melting point of lead (327°C) the solubility of nickel is 0.2 wt.%. The following values for the solubility of nickel in liquid lead have been obtained [109]: 0.153, 0.264, 0.386, 0.465, 0.85, and 1.425 wt.% at temperatures of 370, 450, 500, 540, 635, and 727°C, respectively. At the peritectic temperature the solubility of nickel in lead is approximately 0.11 wt.%.

In contrast to the systems examined earlier, the Ni—Pb system has the following characteristics (Fig. 42): a comparatively low sintering temperature, poor wettability, low solubility of the components, and a lower surface tension of the low-melting component. It can be assumed that shrinkage during sintering in this system will be slight.

We investigated the densification process in samples of the Ni—Pb system, using the apparatus described in Chapter 3 (see Fig. 17). The samples, 52.5 × 11.5 × 11.0 mm (±0.1 mm) in size, were prepared from reduced powders by double-ended pressing with a stop.

The sintering kinetics of Ni—Pb samples was investigated under isothermal conditions in hydrogen at temperatures of 580, 660, 750, and 850°C for 90 min on samples of the following compositions: Ni, Ni + 10 vol.% Pb, Ni + 18 vol.% Pb, and Ni + 28 vol.% Pb. The porosity of all samples was 30%.

Effect of Temperature. Figure 43 shows the shrinkage curves of Ni—Pb samples with different concentrations of lead sintered at different temperatures.

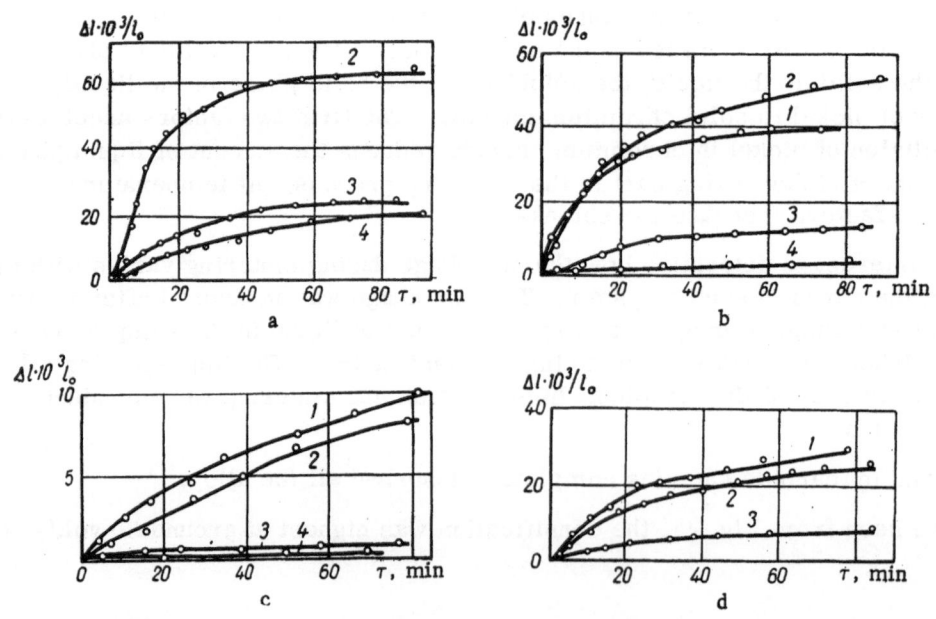

Fig. 43. Shrinkage of Ni—Pb samples in relation to sintering time at different temperatures. a) 850°C; b) 750°C; c) 660°C; d) 580°C. 1) 28 vol.% Pb; 2) 18 vol.% Pb; 3) 10 vol.% Pb; 4) Ni.

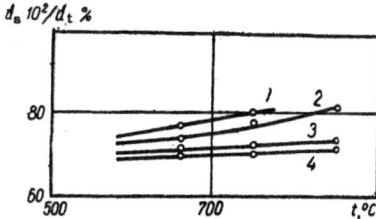

Fig. 44. Density of Ni — Pb samples in relation to sintering temperature and sample composition (sintered for 1.5 h). 1) 28 vol.% Pb; 2) 18 vol.% Pb; 3) 10 vol.% Pb; 4) Ni.

As can be seen from the figure, the shrinkage increases with rise in sintering temperature and with increasing amount of liquid phase in the sample during sintering.

Effect of the Amount of Liquid Phase. In Ni + 28 vol.% Pb samples sintered at 850°C considerable sweating of the liquid phase from the samples was observed, and therefore the variation of the shrinkage could not be plotted at this temperature. At 750°C no sweating of the liquid phase was observed, although the sample contained enough of it for some deformation to be observed as well as slight bulging of the sides of the samples. No sweating of the liquid phase or deformation of the samples was observed at low lead concentrations. The shrinkage increases uniformly with rise in lead concentration and sintering temperature. The increase in the shrinkage of pure nickel with increasing temperature is due to solid-phase sintering of the nickel.

The variation of d_s/d_t with temperature, shown in Fig. 44, indicates that the density of sintered samples increases with rise in temperature and lead concentration. However, for the compositions and temperatures investigated the density did not exceed 82% of the theoretical density.

The Microstructure. Comparison of the microstructures (Fig. 45) of samples sintered at different temperatures shows that an increase of temperature induces sharp grain growth in the refractory phase. Thus, at 850°C the grains are 10 times larger than the grains formed at 580°C and are mainly of irregular shape with rounded edges. It can also be seen that the shapes of the grains are adjusted to each other.

5. Cobalt—Tungsten Carbide System

The sintering of cobalt—tungsten carbide samples also occurs in the presence of liquid phase, the solid phase (WC) being partially dissolved in the liquid.

A phase diagram of the Co—WC system constructed from the data in several publications [45] is shown in Fig. 46.

In the literature there are only few data on the kinetics of densification during the sintering of the Co—WC system. Some information on the densification process in this system can be found in [117], describing a work in which tungsten carbide compacts with 6 and 10% Co were sintered and the length of the samples was measured during heating from 1000 to 1400°C at a rate of about 3°C/min. It was found that in samples of both compositions shrinkage begins at a temperature of about 1150°C and ceases at approximately 1320°C, i.e., at the temperature where the liquid phase appears.

Figure 47 shows shrinkage curves for samples of the Co—WC system (6 and 11% Co), beginning at 1200°C , plotted from the data in [27]. It can be seen from these data that the shrinkage is considerable up to the temperature at which the liquid phase appears.

Cech [64] studied the kinetics of sintering in samples of WC with 15% Co at 1250 and 1350°C in a vacuum and found that at 1250°C, i.e., before the appearance of the liquid phase, the shrinkage was slight (around 10%) during sintering for 3 h. At the temperature of liquid-phase appearance (1350°C) shrinkage was very rapid, practically complete densification occurring in 3-5 min.

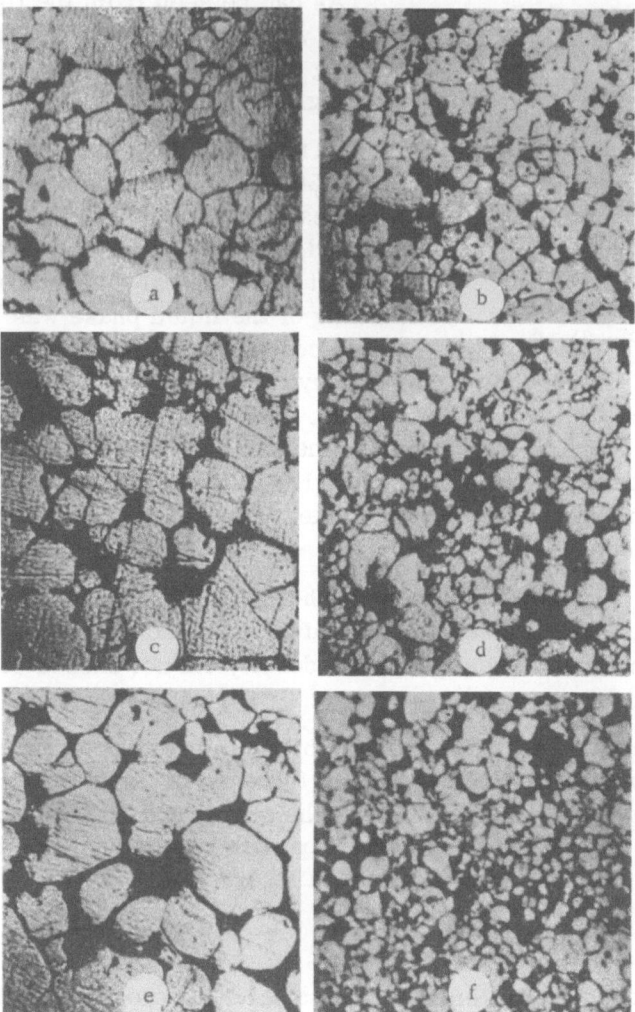

Fig. 45. Microstructure of Ni–Pb alloys sintered for 1.5 h at
850°C (a,c), 750°C (e), 650°C (b,d), and 550°C (f). a,b) Ni + 10
vol.% Pb; c,d) Ni + 18 vol.% Pb; e,f) Ni + 28 vol.% Pb. × 450.

Gurland [76] also investigated the sintering process in the WC–Co system. He studied
the contacts between grains and the extent of their development. The experiments showed that
the formation of a rigid skeleton is possible in the last stage of sintering, although the degree
of its development depends on the cobalt concentration and the sintering time and temperature.
The stage in which the rigid skeleton is formed is characterized by an increase of the carbide
grain size.

Summing up the information on the sintering process in the WC–Co system, we may
note the following:

1. The shrinkage of the porous body begins with solid-phase sintering, which actively
occurs at 1150–1300°C and is accompanied by the formation of a cobalt-base solid solution.

2. When the eutectic temperature is attained and, consequently, the liquid phase appears,
the shrinkage increases sharply and densification becomes rapid. In samples with 25–35 vol.%
Co practically complete densification is attained in 3–5 min.

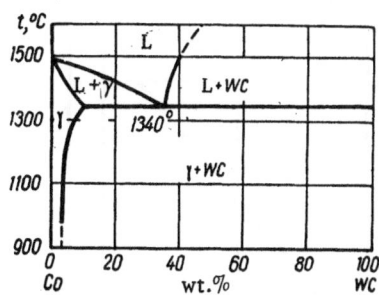

Fig. 46. Phase diagram of Co−WC system.

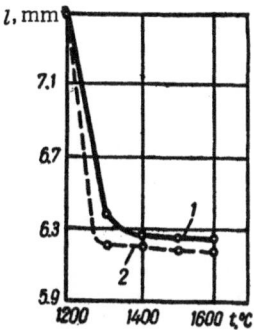

Fig. 47. Shrinkage as a function of sintering temperature for WC + 6 vol.% Co (1) and WC + 11 vol.% Co (2) samples.

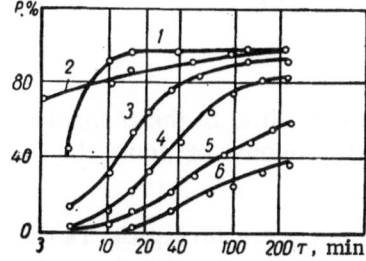

Fig. 48. Densification of Fe−Cu samples in relation to sintering time. 1) 22.0 wt.% Cu, 3 μ; 2) 38.0 wt.% Cu, 9.4 μ; 3) 22.0 wt.% Cu, 9.4 μ; 4) 22.0 wt.% Cu, 15.8 μ; 5) 11.3 wt.% Cu, 9.4 μ; 6) 22.0 wt.% Cu, 33.1 μ.

3. With increase in sintering temperature or time, grain growth of the carbide phase occurs, mainly as a result of recrystallization through the liquid phase.

Grain growth of the carbide phase during sintering is affected substantially by the presence in the mixture of small amounts of other carbides. As investigations have shown, additions of carbides such as vanadium carbide can inhibit the grain growth of tungsten carbide during sintering, which makes it possible to control the grain size of hard alloys.

6. Iron−Copper System

The iron−copper system has been investigated in great detail. Cannon and Lenel have noted that densification occurs already in the first 20 min. The degree of densification depends on the amount of copper, i.e., on the amount of liquid phase, which points to the regrouping mechanism.

Kingery and Narasimhan experimentally determined the degree of densification by measuring linear shrinkage during sintering in dry hydrogen at 1150°C [86]. Samples were prepared from an electrolytic copper powder (325 mesh) and an iron powder with particle sizes of 3, 10, 20, and 40 μ.

Effect of Time. A diagram of densification as a function of sintering time is shown in Fig. 48 and in logarithmic coordinates in Fig. 49. As can be seen from Fig. 49, plots in logarithmic coordinates exhibit two sections. The first fits the relationship $\Delta l/l_0 \sim \tau^k$ (k = 1.3-1.4), which corresponds to the regrouping process, and the second $\Delta l/l_0 \sim \tau^{1/3}$, which corresponds to the solution-and-precipitation process.

Effect of Temperature. The results obtained at temperatures of 1120 and 1200°C indicate that with increasing temperature the degree of densification during sintering increases.

Effect of the Concentration of Liquid. Kingery and Narasimhan [86] have shown that in the presence of a large amount of liquid complete densification can be attained as a result of the regrouping process alone.

Samples with different liquid-phase concentrations were sintered for 4 h and their microstructure was examined. When the concentration of liquid phase at the sintering temperature was 35 vol.% or more,

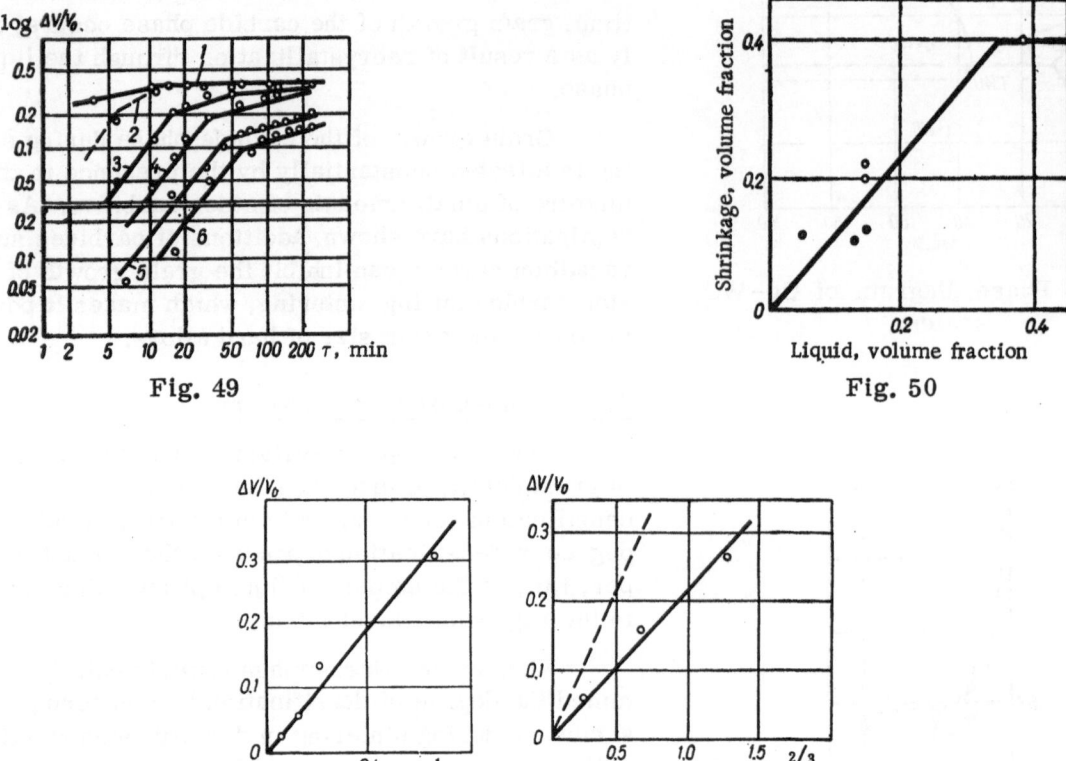

Fig. 49

Fig. 50

Fig. 51

Fig. 49. Densification curves of Fe−Cu samples in logarithmic coordinates. 1) 43 wt.% Cu, 9.4 μ; 2) 22.0 wt.% Cu, 3 μ; 3) 22 wt.% Cu, 9.4 μ; 4) 22 wt.% Cu, 15.8 μ; 5) 11.3 wt.% Cu, 9.4 μ; 6) 22.0 wt.% Cu, 33.1 μ.

Fig. 50. Shrinkage due to the regrouping process in relation to the amount of liquid phase in Fe−Cu samples.

Fig. 51. Variation of densification with iron particle size in samples of the Fe−Cu system. a) Regrouping process; b) solution-and-precipitation process.

complete densification occurred. As can be seen from Fig. 50 (the solid line is the theoretical variation [85] and the circles are the experimental results), at small amounts of liquid the shrinkage ceases long before complete densification is achieved.

Effect of Particle Size. In the initial stage of sintering, i.e., during the regrouping of the particles, the degree of densification is inversely proportional to the particle size (Fig. 51a), while in the solution-and-precipitation stage the degree of densification is inversely proportional to the particle radius raised to the power of $\frac{4}{3}$ (Fig. 51b). The dashed line in Fig. 51b shows the theoretical variation.

Cannon and Lenel did not find such a relationship, although they too noted that the particle size of the refractory component has a considerable effect on densification at the beginning of sintering. Thus, experiments on samples of the Fe−Cu system showed that the degree of densification at a given sintering time is the greater the finer the original iron powder.

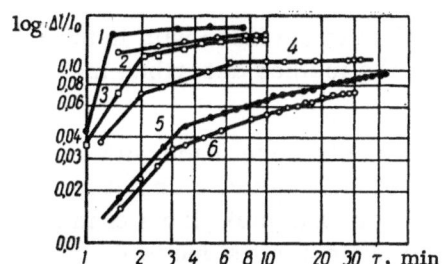

Fig. 52. Densification of MgO—kaolin samples in relation to sintering time at particle sizes of 0.5 μ (4,5,6), 1 μ (3), and 3 μ (1,2) and kaolin concentrations of 2 vol.% (2,3,6), 4 vol.% (1,5), and 8 vol.% (4).

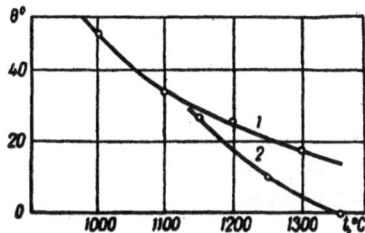

Fig. 53. Variation of angle of contact with temperature for silver (1) and copper (2) on tungsten.

7. Other Systems

Kingery and co-workers [87] investigated a series of nonmetallic systems with limited solubility of the components in each other. The laws governing sintering in these systems have many common features and can be extended to metallic systems. The degree of densification was determined by the method described in [86].

$CoO - B_2O_3$ System. At temperatures above 1250°C the compound CoO is in equilibrium with a liquid phase consisting of approximately 85 vol.% CoO and 15 vol.% Be_2O_3. The microstructure of the alloy containing 20 vol.% of liquid and sintered at 1250°C in helium for 2 h indicates that spherical grains are formed, evidently due to the presence of the liquid, which penetrates the grain boundaries of the solid phase. A similar structure was observed in samples containing 40 vol.% of liquid phase during sintering.

$CaF_2 - NaF$ System. This is a system with a eutectic [97]. Samples of pure CaF_2 with additions of NaF were sintered at 850°C in air, which resulted in a slight densification of the samples. The liquid phase does not completely wet the grains of the solid phase. In $CaF_2 - NaF$ samples however, grain growth is considerable by comparison with pure CaF_2. This system illustrates the necessity for adequate wetting together with solubility and low viscosity of the liquid phase if considerable densification is to be attained.

Forsterite—Talc and Forsterite—Kaolin Systems. Several forsterite-talc and forsterite—kaolin mixtures were prepared. The samples with kaolin were sintered at 1450 and 1550°C to obtain different concentrations of liquid, in accordance with the phase diagram of the $MgO-Al_2O_3-SiO_2$ system [97], while the samples with talc were sintered at 1600°C. High densification of the samples was observed in all cases. The microstructure of both systems is characterized by the formation of forsterite crystals of prismatic shape distributed in the liquid phase at the sintering temperature.

MgO—Kaolin System. Samples were prepared by sintering mixtures of MgO and kaolin in air, in an inert atmosphere, or in a vacuum. They were heated from 1500 to 1750°C in 2 min. Spherical grains in a matrix wetting the boundaries were observed in all cases. The samples were rapidly densified and considerable grain growth occurred. The degree of densification was determined for fine, medium, and coarse MgO powders with 2, 4, and 8% kaolin (4.6, 9.3, and 18.6 vol.% of liquid) at 1750°C. The liquid phase appeared at a temperature of about 1500°C. The results (Fig. 52) indicate three stages of the process. The slope of the straight lines in logarithmic coordinates is equal to unity in the first stage and one-third in the second stage. The degree of densification increases with decreasing particle size of the solid phase and with increasing amount of liquid phase at the sintering temperature. At a large amount of liquid phase (over 35 vol.%) complete densification is attained in the first stage of sintering.

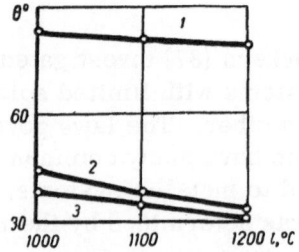

Fig. 54. Variation of angle of contact with temperature for liquid silver (1), Ag + 5 wt.% CuO (2), and Ag + 10 wt.% CuO (3) on aluminum oxide.

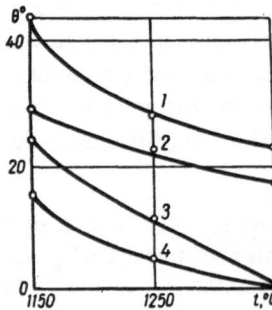

Fig. 55. Variation of angle of contact with temperature for copper (1), Cu + 2.5 vol.% Ni (2), Cu + 6 vol.% Ni (3), and Cu + 10 vol.% Ni (4) on chromium carbide.

8. Wetting

The wetting of solid surfaces with liquid metals was investigated by the method described in Para. 4 of Chapter 3.

Wetting experiments were made with the use of dense substrates of cast metallic or hot pressed (chromium carbide, for example) plates of minimum porosity in order to avoid errors due to the possibility of impregnation.

Wetting of Tungsten. Experiments to determine the wettability of tungsten with liquid copper and silver were made on commercial sheet tungsten 0.5 mm thick. The plates were carefully washed and degreased before the experiments. The copper and silver were pressed into samples of small size.

As can be seen from Fig. 53, the wetting of tungsten improves with increasing temperature. At 1350°C the angle of contact is zero. Copper covers the surface of tungsten completely, coating the substrate with a thin layer not only on the side with the drop but also the reverse side.

The angle of contact of silver on tungsten decreases with increasing temperature. However, it does not reach zero even at 1300°C.

Wetting of Aluminum Oxide. As is well known, pure silver wets aluminum oxide poorly. To improve the wettability, from 5 to 10 wt.% copper oxide was added to silver. An aluminum oxide backing plate, sintered to eliminate open porosity and recrystallized, was used in the experiments.

The experiments were made in air in a horizontal tubular furnace. Direct observations showed that the drop shape changed only during the first minutes.

The results of the investigation (Fig. 54) indicate that pure silver wets aluminum oxide poorly, the angle of contact being only 80° and not changing with increasing temperature. The addition of up to 5 wt.% cupric oxide to silver substantially reduces the angle of contact. Further increase of cupric oxide concentration in silver to 10 wt.% has only little effect on the angle of contact, and the same happens when the temperature is raised from 1000 to 1200°C (the angle changes by about 8-10°).

Wetting of Chromium Carbide. The wetting of chromium carbide with copper and copper-nickel alloys was investigated on chromium carbide samples obtained by hot pressing at 1600°C (maximum residual porosity 5%).

As can be seen from Fig. 55, the angle of contact decreases with increasing temperature. When nickel is added to the alloy the angle of contact decreases considerably. At 6% Ni the angle of contact is zero at a temperature of 1350°C and the alloy completely covers the surface of the chromium carbide.

According to Parikh and Humenik [108], the angle of contact between nickel and titanium carbide in hydrogen is 17°. According to the same authors [82], the angle of contact between cobalt and titanium carbide is close to zero at a temperature of about 1500°C.

It has been reported in the literature that tungsten carbide is completely wetted with liquid cobalt.

In a study of the wetting of nickel with lead [14] it was found that in hydrogen the angle of contact is about 50° and remains practically unchanged with increasing temperature.

CHAPTER 6

SINTERING CHARACTERISTICS OF SYSTEMS WITH SUBSTANTIAL MUTUAL SOLUBILITY OF THE COMPONENTS

There are a number of systems in which it is difficult to distinguish separate stages of sintering. Among these are systems with substantial mutual solubility of the components. The densification process in such systems is a particular case of sintering in the presence of the liquid phase, the mechanism of structure formation having its own particular characteristics. Examples of such systems are Cu–Ni, Cu–Au, W–Mo, Ag–Cd, Cu–Sn, Mo_3Si–V_3Si, Cu–Al, $TaSi_2$–$CrSi_2$, etc.

The Cu–Sn–C system is among those in which the sintering mechanism is not yet fully understood. Alloys of this system are of practical importance, since they are widely used in the electrotechnical and machine-building industries as brush and antifriction materials.

It should be noted that in such systems the liquid phase is not present during the whole sintering time. Due to reactive diffusion during sintering, refractory phases are formed, the liquid phase disappears, and further sintering occurs by the solid-phase mechanism. Studies of the characteristics of phase transformations during the sintering of copper–tin mixtures (90:10) [24] have shown that, once the tin has melted, there is intensive interdiffusion of the components, as a result of which intermetallic compounds are formed — η phase (Cu_6Sn_5), ε phase (Cu_3Sn), δ phase, and finally a copper-base α solid solution. On slow heating the process can occur without the formation of liquid phase, while on rapid heating a liquid phase does appear and takes part in densification. However, the liquid phase is rapidly used up in the formation of more refractory phases.

Investigations by other authors [1, 71] have shown that during the sintering of copper, tin, and graphite samples the area of interparticle contact increases as a result of the interdiffusion of copper and tin. Dilatometric analysis shows that a liquid phase is formed in the course of sintering.

During the sintering of Cu–Sn samples (90:10) [71], considerable sample growth is noted. This may be due to a number of factors — the properties of copper and tin powders, graphite additions, the initial sample density, the time, temperature, and rate of heating during sintering.

Duwer and Martens [68] studied the sintering of the Cu–Ni and Cu–Zn systems, noting that in the former case (Cu–Ni) an equilibrium solid solution is obtained, which is formed as a result of interdiffusion of the components. When samples of the Cu–Zn system are sintered in the temperature range of 160-400°C a considerable increase in sample size occurs. The temperature range in which this effect is observed is independent of the compaction pressure and the quantity of zinc. X-ray analysis has shown that this phenomenon is due to intense diffusion of zinc into copper. Swelling of samples during sintering is not observed if the components have approximately equal diffusion coefficients.

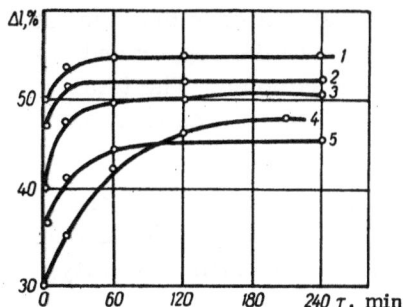

Fig. 56. Variation of shrinkage with sintering time for samples with high initial porosity. 1) Cu + 3 wt.% Bi; 2) Cu + 3 wt.% Pb; 3) Cu + 3 wt.% Sb; 4) Cu; 5) Cu + 3 wt.% Sn.

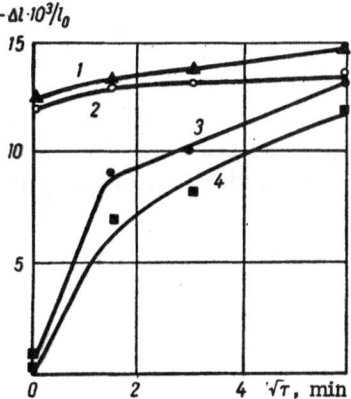

Fig. 57. Variation of linear dimensions with aluminum particle size during sintering of Cu + 10 vol.% Al samples. 1) Below 5 μ; 2) below 50 μ; 3) 100–160 μ; 4) 250–315 μ.

Broquet and co-workers [56, 57] investigated the sintering of copper with 10 and 30 vol.% Ni, 20 vol.% Ag, and 7 vol.% Sn to obtain parts with the maximum density. It was found that the use of alloy powders makes it possible to accelerate sintering, and obtain more uniform shrinkage and higher density. It was assumed that the liquid formed during heating first fills the fine pores, although not instantaneously, and therefore the density of the compact increases gradually with increase in sintering time. Homogenizing annealing is accompanied by the formation of vacancy porosity. An increase of the homogenizing temperature from 850 to 1000°C leads to coalescence of the pores, and during resintering at 1070°C vacancy porosity disappears.

Interesting results were obtained in sintering samples of the Cu–Ag system. In this case there is intensive grain growth, which is evidently due to the solution-and-precipitation process.

Pines [40], studying the effect of the low-melting additions of lead, tin, antimony, and zinc on the sintering of copper, considers that all the additions accelerate shrinkage (in the absence of closed pores) by promoting diffusion of vacancies to the outer surface of the compact in consequence of a decrease of the "potential barrier" for the surface migration of vacancies. This is due to the reduction of the surface tension at the solid–gas interface resulting from adsorption of the low-melting component.

It was found that sintering rate is proportional to the product of the surface tension coefficient and the diffusion coefficient. Although the surface tension coefficient decreases with increasing temperature, the increase of the diffusion coefficient more than makes up for this decrease, and the sintering rate increases. It can be seen from Fig. 56 that all additions accelerate sintering. Closed pores fill up as a result of intergranular vacancy diffusion if there is an adsorption layer of the low-melting addition between the separate grains. In samples with low-melting additions recrystallization rate increases. The linear dimensions of grains in samples of copper with 5 wt.% Pb are 10 times larger than in samples of pure copper at the same sintering temperature and time. This is due to the presence of a liquid phase.

The changes occurring in copper–aluminum samples during sintering in a wide range of temperatures [25] indicate that sintering at a temperature above the eutectic occurs in the presence of a liquid phase. An alloy of copper and aluminum is formed as a result of diffusion of aluminum from the liquid phase into the solid copper. On slow heating the liquid phase is formed at the contacting surfaces in small quantities and conditions required for the regrouping of particles are not established. At a high heating rate a large quantity of liquid forms in a short time. In this case a regrouping of particles occurs.

Since the diffusion coefficient of aluminum in copper is considerably higher than that of copper in aluminum, the result of diffusion is that aluminum migrates into the copper and voids form in the aluminum particles, causing considerable expansion of the samples. The total change in compact size is due to these two factors — liquid-phase shrinkage and an increase in volume.

The increase in volume depends not only on sintering time but also on the size of the aluminum particles (Fig. 57). The smaller the aluminum particle size, the larger the area of contact between the copper and aluminum particles in compacts at the same initial porosity.

The results of the investigation indicate that in this case, too, the liquid phase plays a substantial role in alloy formation, since the liquid phase is responsible for the appearance of compressive forces drawing the particles together and causing their rearrangement.

The further course of the shrinkage or expansion processes depends on the relative magnitudes of the diffusion coefficients of the components. In those cases where the diffusion coefficient of the low-melting component is higher than that of the high-melting component, sample growth is observed. If the diffusion coefficients are practically equal, no expansion occurs and on accelerated heating the samples may shrink. When the diffusion coefficient of the low-melting component is lower than that of the refractory component, shrinkage would be expected to occur.

The sintering processes in systems with substantial mutual solubility of the components has not been studied sufficiently. Investigations are needed of such important factors as sintering temperature, time, initial porosity, heating rate, and concentration and particle size. Further detailed studies of similar systems are required to clarify the role and effect of the liquid phase on sintering processes in these systems.

CHAPTER 7

MODELING OF CAPILLARY FORCES ACTING DURING SINTERING IN THE PRESENCE OF A LIQUID PHASE

Modeling is of considerable importance in theoretical studies of the solid-phase sintering mechanism. Geguzin [7], Pines [40], Cabrera [58], Kingery [88], and Kuczynski [89] have made substantial contributions to the theory of these processes by modeling solid-phase sintering. Elementary models have hardly been used to study the mechanism of liquid-phase sintering. The sample being sintered in the presence of a liquid phase is a dispersed capillary system. The elements of such a system are composed of two solid particles with a layer of liquid phase at the contact between them. The presence of the liquid and adequate wetting of the particle surface induces compressive capillary forces that draw the particles together. This force is responsible for shrinkage during sintering. Depending on the size of the particles and the surface tension of the liquid, the capillary pressure can reach (for liquid metals) tens and hundreds of daN/cm^2. In theoretical calculations the pressure p is usually taken as equal to σ_L/r, where σ_L is the surface tension of the liquid metal phase and r is the pore radius, which is of the same order as the particle radius.

This relationship is a rough estimation and even qualitatively cannot express the various functional relationships (the pressure as a function of the quantity of liquid phase, wettability, distance between particles, etc.). A detailed investigation of such relationships in an elementary model of a sintered body — a pair of particles with liquid at the contact between them — was undertaken in [37-39].

Between two particles with a dry contact molecular bonding forces operate. A general expression for these forces was obtained by Deryagin [11]. For two spheres of radii r_1 and r_2 the force is given by the relationship

$$F = 2\pi \frac{r_1 r_2}{r_1 + r_2} (2\sigma_{1-3} - \sigma_{1-1}), \qquad (47')$$

where σ_{1-3} is the free energy per unit area of the interface between one sphere and an intermediate phase 3 separating it from the other sphere; σ_{1-1} is the free energy per unit area of the interface between the two spheres.

The presence of liquid phase at the particle contact, forming a curved interface and three-phase wetting perimeters, induces capillary forces. A comparison of the magnitudes of these two groups of forces indicates that the forces occurring owing to the presence of liquid at the point of contact are considerably larger (by two to three orders) than the molecular forces.

The first suggestion that solid particles are drawn together by the surface tension of a thin film of liquid between them was made by Stone [118], although he gave no quantitative calculations of the forces. McFarlane and Tabor [72] later showed that for the case of a spheri-

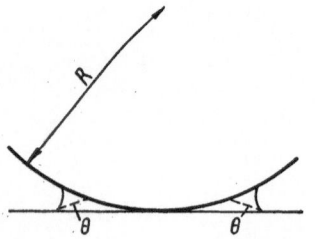

Fig. 58. Diagram of a spherical particle on a plane surface.

cal particle on a plane surface (Fig. 58) with a thin film of liquid between them the cohesive force due to surface tension can be expressed by

$$F = 4\pi R\sigma \cos\theta, \qquad (48)$$

where F is the cohesive force, R is the sphere radius, σ is the surface tension of the liquid, and θ is the angle of contact.

An experimental check showed that this equation is applicable when the thickness of the liquid film and the angle of contact are small.

Larsen [46] used a centrifuge to measure the adhesion between glass spheres with R = 67 and 84 μ and a plate in the presence of a liquid film and verified Eq. (48). He also derived the following expression for the adhesion between spheres and a cylinder in the presence of a liquid film:

$$F = 2\pi R\sigma \left\{ \frac{\rho/R}{[(\rho/R)^2 + (r/R)^2]^{1/2}} + \frac{1}{[(\rho/R)^2 + 1]^{1/2}} \right\} (1 + R/r), \qquad (49)$$

where r is the cylinder radius, ρ is the radius of the meniscus of the entrapped liquid, and R is the sphere radius.

In experiments with vibrating glass filaments (r = 128 and 420 μ) and spheres (R = 55 μ) lubricated with oil, Larsen found that the spheres begin to separate when the maximum inertia $ma(2\pi\nu)^2$ (a is the amplitude and ν the frequency of vibrations, m is the mass of the sphere) is somewhat larger than the adhesive force calculated by Eq. (49).

Davies [69] calculated the capillary effect for two solid spherical particles, taking into account the amount of entrapped liquid and the angle of contact between the liquid and the solid surface on the basis of the Laplacian pressure drop. No experimental check of these calculations was made. The Laplacian relationship was used in [5] to calculate the forces acting between particles.

The thermodynamic theory of adhesion developed by Deryagin can also be used for the case where a third phase is held in the zone of contact of solid surfaces by capillary forces. The presence of this phase, if it is strongly lyophilic with the material of the solid particles, substantially increases the adhesive force in this case [11].

Pokrovskii [42] investigated capillary forces in soils and determined the effect of the quantity of entrapped liquid and the shape of the contact on the magnitude of the cohesive force with complete wetting of solid particles by the liquid. He derived a formula relating the quantity of entrapped liquid to the magnitude of the cohesive force

$$F = \pi\rho_1\sigma \left(\frac{1}{\rho_2/\rho_1} - 1 \right), \qquad (50)$$

where ρ_1 and ρ_2 are the radii of curvature of the liquid (ρ_1 and ρ_2 depend on the amount of entrapped liquid).

Thus, at various times, different authors have investigated the cohesive force between solid particles both for a small amount of liquid and without taking its effect into account [11, 72] and for the case of complete wetting [42]. No complete analysis of these forces in relation to different conditions or their experimental verification have been performed.

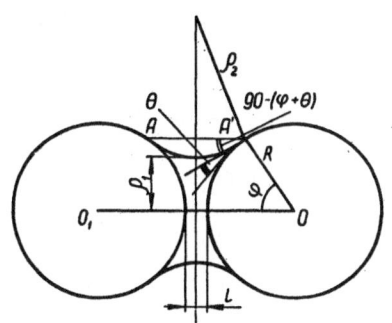

Fig. 59. Diagram of the contact between two spherical particles with an intermediate liquid film.

Complete wetting is relatively rare in the practice of liquid-phase sintering of metallic systems and in other metallurgical systems. It is therefore of interest to determine the effect of the degree of wetting, the amount of liquid phase, the shape of the particles, and other parameters of the dispersed system on the magnitude of the capillary forces.

Since in practice one frequently encounters capillary systems in which the particles are nonspherical – e.g., cubes, prisms, etc. – the characteristic type of contact in this case will be between the corner of a cube or prism and a plane face. According to Pokrovskii [42], this type of contact can be approximated by the contact of the apex of a cone with a plane. Not infrequently capillary systems consist of spherical particles alone (for example, filter materials). Therefore, the contact of a pair of spherical particles and the contact of the apex of a cone with a plane were used as the basic models of a capillary system. We have calculated the compressive capillary forces and analyzed them in relation to various factors, and also experimentally verified the equations obtained.

1. Spherical Particles

Calculation of the Contracting Forces.   The capillary force drawing two particles together must consist of two components.

1. Due to the curvature of the meniscus on both sides of the surface of the liquid there will be a pressure differential, determined by Laplace's first law, which for the given case is

$$\Delta p = \sigma_L \left(\frac{1}{\rho_2} - \frac{1}{\rho_1} \right),   \tag{51}$$

where σ_L and ρ_1, ρ_2 are, respectively, the surface tension and radius of curvature of the surface of the liquid cup (Fig. 59).

Within the liquid the pressure is lower than outside it; the particles are subjected to a compressive force

$$F_1 = \Delta p S,   \tag{52}$$

where S is the area of the projection of the liquid–solid interface onto a plane perpendicular to the direction of the force.

2. It is assumed that each solid particle in the pair under consideration is partially immersed in the liquid. Along the contact perimeter of the liquid and the solid particle acts a force determined by the projection of the surface-tension vector onto the line OO_1 (see Fig. 59). The second component of the compressive force for two particles is

$$F_2 = c\sigma_L \sin(\varphi + \theta),   \tag{53}$$

where c is the length of the wetting perimeter, φ is an angle depending on the amount of liquid, and θ is the angle of contact.

Attention should be drawn to the fact that in earlier investigations the second component of the cohesive force was not taken into account even though, with a sufficient amount of liquid, this component makes a substantial contribution to the capillary force.*

----

*This question has been raised in some recent publications [124, 125].

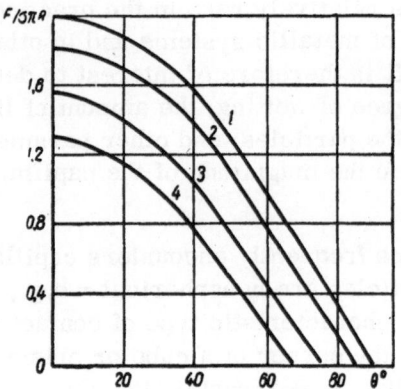

Fig. 60. Variation of cohesive force with the angle of contact at: 1) $\varphi = 0°$; 2) $\varphi = 10°$; 3) $\varphi = 30°$; 4) $\varphi = 50°$.

Thus, the total force drawing two spherical particles together will be

$$F = F_1 + F_2 = \sigma\left[\pi R^2 \sin^2 \varphi\left(\frac{1}{\rho_2} - \frac{1}{\rho_1}\right)\right.$$
$$\left. + 2\pi R \sin \varphi \sin(\varphi + \theta)\right], \tag{54}$$

where

$$\rho_1 = R \sin \varphi - \left[R(1 - \cos \varphi) + \frac{l}{2}\right]\frac{1 - \sin(\varphi + \theta)}{\cos(\varphi + \theta)}; \tag{55}$$

$$\rho_2 = \frac{R(1 - \cos \varphi) + \frac{l}{2}}{\cos(\varphi + \theta)} \tag{56}$$

on the assumption that AA^1 is a circular arc with a radius ρ_2 and l is the gap between the particles (see Fig. 59).

With complete wetting, i.e., when $\theta = 0$ and $l = 0$, Eq. (54) takes the following form:

$$F = \sigma\left[\pi R^2 \sin^2 \varphi\left(\frac{1}{\rho_2} - \frac{1}{\rho_1}\right) + 2\pi R \sin^2 \varphi\right], \tag{57}$$

where

$$\rho_1 = R\left[\sin \varphi - (1 - \cos \varphi)\frac{1 - \sin \varphi}{\cos \varphi}\right];$$
$$\rho_2 = R\frac{1 - \cos \varphi}{\cos \varphi}.$$

The volume of the liquid cup at $\theta = 0$ is

$$V = 2\pi R^3 (\sec \varphi - 1)^2\left[1 - \left(\frac{\pi}{2} - \varphi\right)\tan\varphi\right]. \tag{58}$$

The formula obtained for the cohesive force does not take into account the effect of the gravitational field, which will change the surface curvature of the liquid between the two particles. Therefore, Eq. (54) holds true either for small angles φ (at small amounts of liquid) or at any values of φ for particles of small size, when the force of gravity is negligible by comparison with the surface forces.

Analysis of Eq. (57) at $l = 0$ indicates that, with increasing amount of liquid, the force drawing the particles together decreases; the contracting force will be highest at $\varphi \to 0$, i.e., when the amount of liquid tends to zero, Eq. (57) in such a case becoming transformed into the well-known formula

$$F = 2\pi R \sigma \cos \theta. \tag{59}$$

Let us consider how the magnitude of the capillary force must be affected by changes in the angle of contact θ at different angles φ. Figure 60 shows the variation of the cohesive force with the angle of contact θ [57] for different amounts of liquid and different values of φ. As the angle of contact deviates from zero the cohesive force sharply decreases and becomes zero at the wetting angle $\theta_{cr} < 90°$.* With further increase of the angle of contact the particles begin to repel each other.

---

*$\theta_{cr} = 90°$ only for $\varphi = 0$, i.e., for an infinitely small amount of liquid.

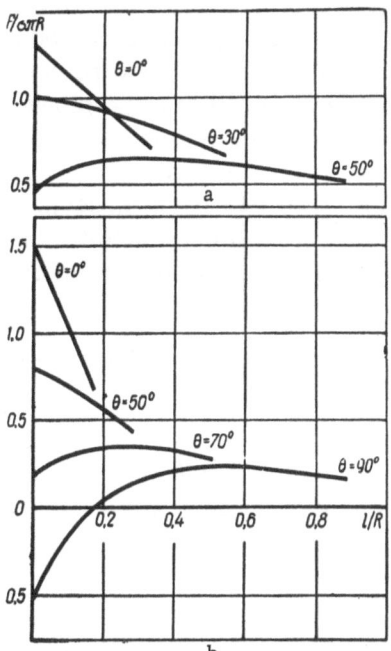

Fig. 61. Variation of cohesive force with particle gap at $\varphi = 50°$ (a) and $30°$ (b).

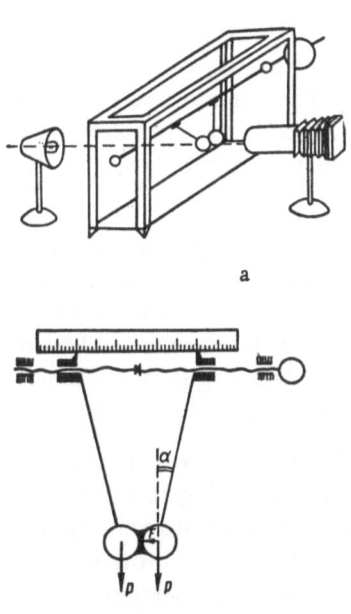

Fig. 62. Diagram of the apparatus for measuring the forces between two particles (a) and diagram of the measurements (b).

From the relationship $F = f(\varphi_1, \theta) = 0$ one can find the values of φ and θ_{cr} at which the particles cease to be drawn together by capillary forces. The solution of this equation is

$$\theta_{cr} = 90 - \frac{\varphi}{2}.$$

This relationship is of fundamental importance. In the region where the sum of the values of θ and $\varphi/2$ does not exceed $90°$ a compressive force operates; otherwise the particles are repelled. For example, in liquid-phase sintering a compressive capillary pressure (leading either to a regrouping of the particles or to the solution of the material of particles in contact, as shown by Kingery [85]), will be absent, i.e., shrinkage and sintering will not occur. If wetting is incomplete, the variation of the cohesive force with particle spacing will be different. When the particles move apart, the parameters φ and l change so that the volume of the liquid cap V remains constant.

The volume of the liquid cup for the case of incomplete wetting is given by

$$\frac{V}{\pi} = 2\left\{\cos(\varphi + \theta) - \left[\frac{\pi}{2} - (\varphi + \theta)\right]\right\}(\rho_2^3 + \rho_1\rho_2^2)$$
$$+ \rho_1^2\rho_2\cos(\varphi + \theta) \tag{60}$$

[the values of ρ_1 and ρ_2 being once again given by Eqs. (55) and (56)].

These relationships were calculated with the aid of Eq. (54). The results are shown in Fig. 61, from which it can be seen that at $\theta = 0$ the cohesive force decreases with increasing l; for small angles θ, an intermediate region is observed where the cohesive force slowly decreases with increasing l. With further increase in the angle of contact, the cohesive force reaches a maximum and finally, for angles $\theta > 90°$, the particles are repelled, i.e., the force becomes negative. But as the gap between the particles increases, the cohesive force increases, passes through zero, attaining a peak in the positive region and then decreasing.

Experimental Part. There are various methods for determining cohesive forces: by separating particles with a spring microbalance [55] or a pendulum [72], by tilting the support plate and rotating it about vertical and horizontal axes, by employing vibrations and applying a blast to a sprayed surface [10, 23, 47], etc.

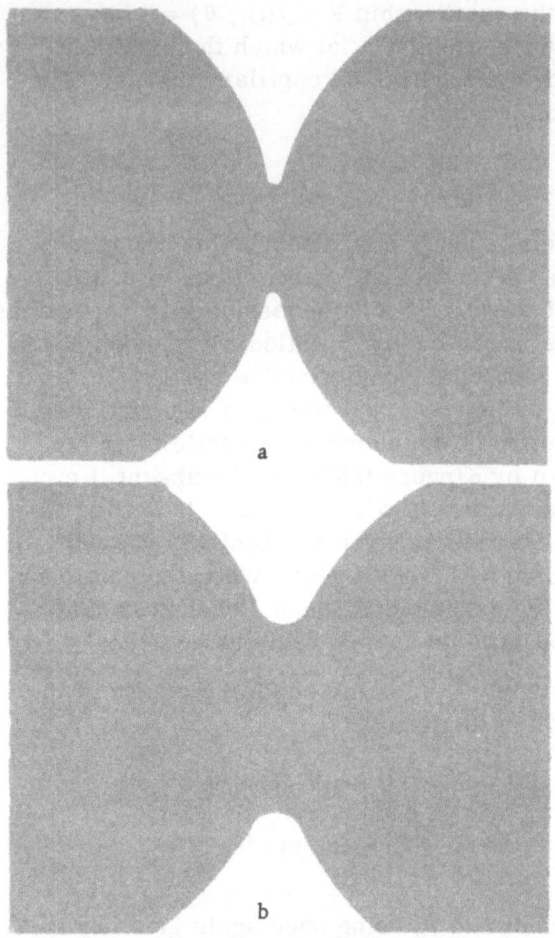

Fig. 63. Contact between two balls with different amounts of liquid. a) 0.027 mm³; b) 0.12 mm³ (steel balls, liquid petrolatum).

To measure the forces between two particles the authors designed and constructed a special apparatus (Fig. 62). Balls were suspended from threads fastened to slide blocks which could be moved apart by turning a microscrew. The blocks and the microscrews were provided with right- and left-hand threads. The distance between the slide blocks (points where the threads were fastened) was measured on a scale .

The force drawing two particles together is given by the equation

$$F = P \tan \alpha = P \frac{\frac{S-2R}{2}}{\sqrt{l^2 - \left(\frac{S-2R}{2}\right)^2}}, \qquad (61)$$

where S is the distance between the slide blocks at the point where contact was broken, and R is the radius of the balls.

By determining the distance between the slide blocks at the point where the balls are separated from each other and knowing the ball weight P, one can calculate the cohesive force F. One can also introduce corrections for the weight of the thread and the weight of the liquid into the equation. The thread used in the experiments was an 80% Ni–20% Cr wire 20 μ in diameter, and its weight was neglected.

The sensitivity of the apparatus can be regulated by changing the weight of the balls and the length of the suspension threads. The entire system was enclosed in a glass-walled box. The image of the balls with the liquid between them was projected onto a screen and a photographic plate; it was photographed at a magnification of 14 times. Measurements were made with a TM telescopic microscope and a Fotokor camera. The angle φ was measured from the photographs and from it was calculated the volume of the liquid cap, the angle of contact θ, and the particle gap. A measuring microscope was used to measure the gap from the photographs.

Measurements were made in the following manner. Dry, clean balls were brought together until they touched and then separated to be sure they were not sticking to each other (absence of residual magnetism or contamination). Then a small amount of liquid was applied between the balls with a special thin probe, causing the spheres to adhere to each other. After this, the screw was turned to move the slide blocks apart to a distance somewhat less than the critical (in the case of complete wetting) or one producing a tiny gap between the balls (in the case of incomplete wetting). The image of the balls with the liquid cup was focused and photographed on a plate. Then the slide blocks were again separated slightly and the image of the balls with the larger gap between them was photographed; the process was repeated until contact was broken. Experiments were made with both larger and smaller amounts of liquid between the balls, the symmetry of the meniscuses of the liquid cup being carefully examined. In

Fig. 64. Cone/plane type of contact between two particles with entrapped liquid.

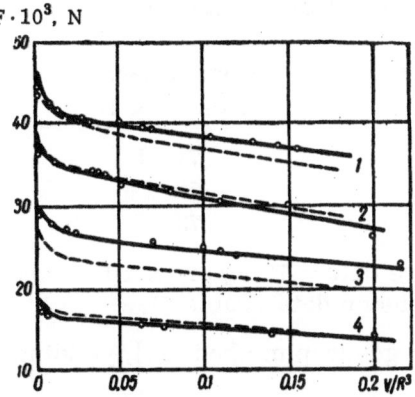

Fig. 65. Variation of contracting force with amount of liquid. 1) Benzyl alcohol (R = 0.2 cm); 2) oil (R = 0.2 cm); 3) ethyl alcohol (R = 0.2 cm); 4) oil (R = 0.1 cm).

most cases the meniscuses of light liquids were symmetrical (Figs. 63 and 64); asymmetry was noted only with large amounts of liquid, where $\varphi = 40$-$50°$.

Effect of the Amount of Liquid Phase on Cohesive Force. The liquid phases used were liquid petrolatum, benzyl and ethyl alcohols, and mercury. With the first three substances steel bearing balls 1 and 2 mm in radius were used. Steel is well wetted by the alcohols and the oil — the angle of contact is close to zero. The weight of the steel balls was 0.259 and 0.032 g, the thread length being 19.80 and 9.60 cm, respectively. The weight of the threads and the liquid was less than 0.001 g (which was neglected). For the experiments with mercury we used gold balls with radii of 0.16 and 0.135 cm, the thread length being 8.90 and 6.80 cm, respectively (a correction was introduced for the weight of the mercury).

The most precise measurements can be performed with the oil and with benzyl alcohol. These substances completely and evenly wet the surface of balls, are nonvolatile, and have a low density and, therefore, the meniscuses of the liquid cup are not distorted.

In experiments with ethyl alcohol measurements are complicated by the intense evaporation of the liquid during the experiment. With mercury the meniscuses are distorted owing to the high density wetting hysteresis, and intense amalgamation of gold, resulting in lower measurement accuracy.

As can be seen from Figs. 65 and 66, the force drawing a pair of spherical particles together decreases with increasing amount of liquid phase. The cohesive force increases with increasing surface tension of the liquid and particle radius.

The magnitude of the force drawing the particles together and its variation with the amount of liquid were also calculated, using Eq. (57), at $\theta = 0$. To do this it was necessary to know the surface ten-

TABLE 7. Surface Tension of the Oil and Alcohols

Substance	Capillary diameter, cm			Average value $N/cm^2 \cdot 10^5$
	0.27	0.33	0.66	
Oil	33.6	33.5	33.4	33.5
Ethyl alcohol	23.6	23.0	22.5	23
Benzyl alcohol	39.3	39.0	—	39

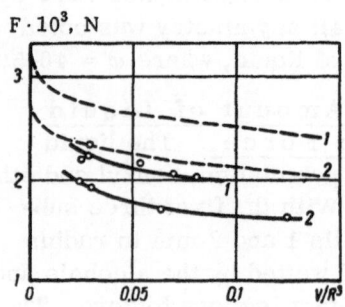

Fig. 66. Variation of contracting force with amount of mercury between balls. 1) R = 0.16 cm; 2) R = 0.13 cm.

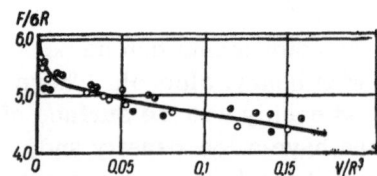

Fig. 67. Generalized curve of contracting force plotted against amount of liquid: ○ Oil (R = 0.2 cm); ◑ benzyl alcohol (R = 0.2 cm); ● oil (R = 0.1 cm).

sion of the liquids investigated. Tabular data do not exist for all the substances investigated and, in addition, surface tension depends greatly on the purity of the liquid, and in view of this the surface tension of the substances investigated was measured experimentally. For the light liquids measurements were made by the drop count method, using glass capillaries (outside diameters 0.33, 0.27, and 0.66 cm). By controlling the flow of the liquid it was possible to slow down the time for the formation of a drop to 5–10 min.

The surface tension was calculated from the equation

$$\sigma = \frac{mg}{r} F\,(V/r^3), \qquad (62)$$

where m is the mass of the drop, r is the radius of the capillary, and $F\,(V/r^3)$ is a function of the drop volume and the capillary radius, determined with the aid of Harkins and Brown's tables [3].

The density of the liquid was measured only for the oil, tabular data being used for the other substances. The apparatus was calibrated by determining the surface tension of distilled water with different capillaries. The following surface tension values were obtained:

Diameter of capillary, cm	Surface tension, N/cm^2 ($\times 10^5$)
0.66	71.11
0.33	72
0.27	73

The results obtained are in good agreement with literature data — 72 N/cm × 10^{-5} — indicating that the instrument gives sufficiently reliable surface-tension data (Table 7).

For mercury the surface tension was measured by the large-drop method [32], yielding a value of 0.00340 N/cm^2.

Using the surface-tension values obtained we plotted the theoretical variation of the contracting force as a function of the amount of liquid (see Figs. 65 and 66, dashed lines).

As can be seen from the figures, the theoretical and experimental data are in good agreement; for benzyl alcohol and oil the discrepancy does not exceed 1–2%. The discrepancy was larger in the case of ethyl alcohol (about 10%) and mercury (about 20%), but these measurements were less precise. In particular, in the case of mercury, error was due to the fact that mercury intensely reacts with and spreads over gold, without forming a sharply defined wetting perimeter (three-phase boundary), so that the value of the second term in Eq. (57) decreases.

Figure 67 shows a generalized curve (solid line) of the contracting force plotted against the amount of liquid, $F/\sigma R = f\,(V/R^3)$.

The values of cohesive force should fit this curve regardless of the surface tension of the liquid (with complete wetting) or the particle radius. Indeed, the points obtained for the systems investigated, having different surface tensions and ball radii, form a single curve (the most pre-

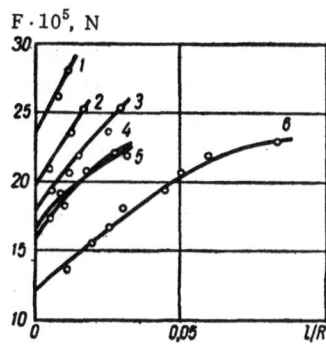

Fig. 68. Variation of cohesive force with size of the gap at different amounts of glycerin (cm³): 1) 0.000175; 2) 0.000627; 3) 0.00082; 4) 0.000952; 5) 0.00115; 6) 0.0023.

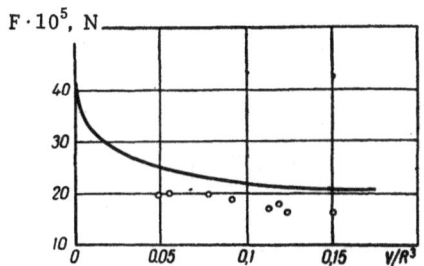

Fig. 69. Variation of cohesive force with amount of liquid.

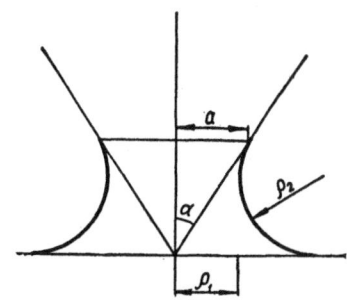

Fig. 70. Diagram of cone–plane contact with entrapped liquid.

cise data, for benzyl alcohol and oil, were plotted). It should be noted also that the ball gap with a liquid was zero whenever the angle of contact was zero; thus, the presence of a liquid cup wetting the balls presses the particles together instead of repelling them, as confirmed by Eq. (57). This is important, since the opposite opinion has been expressed in the literature [5].

<u>Effect of the Degree of Wetting on Cohesive Force.</u> The effect of the degree of wetting was studied with steel balls with a radius of 0.2 cm, weighing 0.259 g. Glycerin, whose angle of contact with a steel surface is different from zero, was used as the liquid phase. The length of the threads was 20.2 cm. The weight of the threads and the liquid was less than 0.001 g, which was neglected.

The angle of contact of the glycerin was determined from photographs. In the calculations we used an average value of $\theta = 55°$.

The value of l measured directly through the film of liquid can be distorted by optical phenomena. For that reason we also measured l on photographs as the difference $(4R + l) - 4R$.

Comparison of the results obtained by both methods showed that they are in satisfactory agreement, i.e., whatever optical distortion occurs is relatively small. On the other hand, there can also be an error in determining l as the difference between two large values.

As can be seen from Fig. 68, the force increases with increasing gap, and at a certain critical gap, when the value of the contracting force reaches a maximum, the balls break apart. By extrapolation of the values obtained to the axis of ordinates we determined the cohesive force at $l = 0$, which should be given by Eq. (54).

Along with the experimental determinations, the cohesive force was calculated by Eq. (54) at $\theta = 55°$. The drop count method was used to find the surface tension of glycerin – 60.1 dyn/cm. The curve in Fig. 69 shows the theoretical variation and the points plot the experimental results. As can be seen, the force drawing the particles together decreases with increasing quantity of liquid phase also at $\theta \neq 0$. The theoretical and experimental values are in fairly good agreement. The slight discrepancy between the values may be due to the difficulty in determining small values of the gap between the particles from photographs, since an inexact value of l can result in an error of several percent in extrapolating to zero.

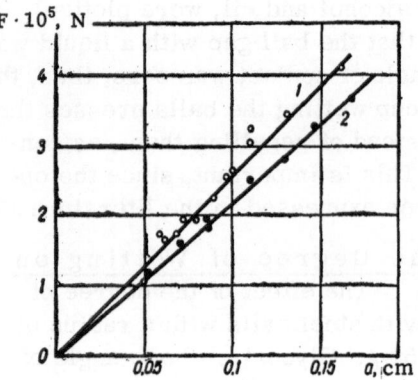

Fig. 71. Variation of cohesive force with the radius of the circular wetting perimeter of the cone. 1) Benzyl alcohol; 2) liquid petrolatum.

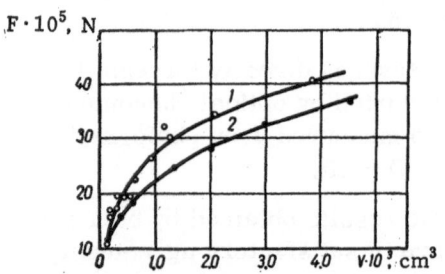

Fig. 72. Variation of contracting force with the volume of liquid. 1) Benzyl alcohol; 2) liquid petrolatum.

2. Nonspherical Particles

For nonspherical particles the total cohesive force (Fig. 70) is the same as for spherical particles, and consists of two components, the first accounting for the Laplace pressure differential on both sides of the surface of the liquid due to the curvature of the meniscuses, and the second for the projection of the surface-tension vector onto the line OO, which acts along the wetting perimeter. The second component is proportional to the length of the wetting perimeter of a conical particle.

Adding these two components, we obtain the following expression for the cohesive force:

$$F = \pi a^2 \sigma \left(\frac{1}{\rho_2} - \frac{1}{\rho_1} \right) + 2\pi a \sigma \cos \alpha, \qquad (63)$$

where σ, ρ_1, and ρ_2 are the surface tension and radii of curvature of the liquid, a is the radius of the circular wetting perimeter of the cone, and α is the angle between the height and the generatrix of the cone.

In this case,

$$\rho_1 = a \left[1 - \frac{1 - \cos \alpha}{\tan \alpha (1 + \sin \alpha)} \right],$$

$$\rho_2 = \frac{a}{\tan \alpha (1 + \sin \alpha)}.$$

After substitution of the known angle α, we find

$$F = Ba,$$

where B is a constant which is independent of the quantity of liquid. The volume of the liquid cup in this case is given by the following relationship:

$$V = \pi \rho_2 \left\{ [(\rho_1 + \rho_2)^2 + \rho_2^2] (1 + \sin \alpha) - \frac{\rho_2^2}{3} (1 + \sin^3 \alpha) - (\rho_1 + \rho_2) \rho_2 \left(\sin \alpha \cos \alpha + \alpha + \frac{\pi}{2} \right) \right\}. \qquad (64)$$

Knowing the angle α, we obtain

$$V = Ca^3,$$

where C is a numerical factor depending on α. Then

$$F = kV^{1/3},$$

where k is a constant.

In a similar way one can obtain an equation for the case of incomplete particle wetting with a liquid when the particle gap is not equal to zero. As analysis of the equation indicates, for a given volume of liquid the force increases with increasing angle α. It also follows from the equation that the force drawing the particles together must increase with increasing volume of liquid (and not decrease, as in the case of spherical particles!). This finding has been confirmed experimentally.

In our experiments the plane surface was the end face of a steel cylinder 4 mm in diameter. The cone had an angle of 96° at the tip, and both the cone and plane surfaces were carefully polished. The weight of the cone and the cylinder was the same (m = 0.448 g). The cone and the cylinder were suspended on 80% Ni–20% Cr alloy threads 20 μ thick so that their centers of gravity coincided with the thread suspension centers to avoid misalignment. The liquids were liquid petrolatum and benzyl alcohol. The value of a was measured from photographs by means of a measuring microscope.

As can be seen from Fig. 71, the variation of F with a is linear, confirming the theoretical relationship F = Ba.

Figure 72 illustrates the variation of the contracting force with the volume of liquid. The force drawing the particles together increases with increasing amount of liquid phase. For benzyl alcohol the curve is higher than for petrolatum because of the greater surface tension of benzyl alcohol. Comparison of the experimental results with the theoretical data shows that they are in close agreement.

Thus, the character of the variation of cohesive force with the amount of liquid differs completely from the case of spherical particles.

The equation obtained can be used to determine the value of the capillary force for particles with any type of contact, provided the geometry changes are taken into account when determining the radii of curvature.

From the results of the theoretical and experimental studies we conclude that:

1. The capillary force of cohesion decreases with increasing amount of liquid between spherical particles and increases with increasing amount of liquid for contacts of the cone–plane type.

2. The cohesive force decreases with increasing angle of contact.

3. The cohesive force increases as the gap between the particles decreases and as the surface tension of the liquid and particle size increase.

We also determined the critical values of angle of contact, at which compressive forces, leading to densification during sintering, still occur.

Comparison of the experimental and theoretical data shows good agreement, the difference not exceeding 1–5%.

GENERAL CONCLUSIONS

On the basis of our experiments and literature data it is possible to draw certain general conclusions regarding liquid-phase sintering processes, and also clarify a number of specific phenomena and rules of sintering.

There are three basic processes controlling shrinkage during liquid-phase sintering.

1. The Regrouping Process.   The liquid formed induces capillary forces between the particles (compressive at a certain degree of wetting of the particles by the liquid). With a relatively small degree of pore filling by the liquid the force drawing a pair of spherical particles together is given by

$$F = \sigma \left[\pi R^2 \sin^2 \varphi \left(\frac{1}{\rho_2} - \frac{1}{\rho_1} \right) + 2\pi R \sin \varphi \sin (\varphi + \theta) \right].$$

The area of the plane surface occupied by one particle is, of course, equal to $4R^2$ (with cubic packing), so that the pressure with which one layer of particles is pressed against a second layer is

$$p = \frac{F}{4R^2} = \sigma \left[\frac{\pi}{4} \sin^2 \varphi \left(\frac{1}{\rho_2} - \frac{1}{\rho_1} \right) + \frac{\pi}{2R} \sin \varphi \sin (\varphi + \theta) \right].$$

But, since ρ_1 and ρ_2 are proportional to R (particle radius) the pressure is inversely proportional to the particle radius. Therefore, the smaller the particle size the greater the densification.

The presence of liquid not only causes capillary pressure which promotes displacement of the particles, but also makes such displacement much easier, since the liquid flows into the space between the particles and acts as a lubricant, thereby sharply reducing the friction and sticking of the particles.

2. Solution-and-Precipitation.   The compressive forces on the particle contacts lead to more intense dissolution of the solid-phase material in these areas and to transfer of the material from the contact zones to the free surface, as a consequence of which the particles are brought closer together and the sample is densified.

3. Formation of a Rigid Skeleton.   With insufficient wetting of the solid-phase particles by the liquid the latter does not penetrate into the joints between the particles and "dry" contact results. In this case the two preceding mechanisms are either entirely absent or strongly inhibited, and shrinkage is small or altogether absent. A rigid skeleton is ordinarily formed in those cases where the solid-phase material exhibits good solid-phase sinterability at temperatures below the melting point of the low-melting component, so that particle welding occurs, σ_{S-S} is small, and the dihedral angle is large.

66

In the first stage of sintering (regrouping of the particles) densification is described by the relationship $\Delta l / l_0 \sim \tau^{1-2}$, i.e., the process is accelerated. This is explained by the fact that the solid-phase particles are separated from each other in the compact by particles of the low-melting component. When a liquid phase appears during sintering the solid particles begin to be drawn together by the capillary forces and the distance between them decreases. Considering the results obtained for capillary cohesive forces in the elementary model, one can see that the compressive forces increase with decreasing distance between the particles (see Fig. 61).

Thus, as a result of the drawing together of the particles during sintering the contracting forces increase, i.e., the process is self-accelerating. The sintering process is also accelerated by the presence of closed pores, the size of which decreases during sintering, while the pressure in them increases.

In studies of the W−Cu, W−Ag, and certain other systems, a sharp increase of shrinkage with rise in temperature was noted. Since the solid phase in these systems does not dissolve in the liquid phase, the most substantial changes with increasing temperature will occur in the viscosity of the liquid phase and, especially, the angle of contact. Obviously, the reasons for the variation of shrinkage with temperature should be sought in these changes.

The surface tension of the liquid changes only negligibly with increasing temperature and can have no substantial effect on the densification process. The decrease of the viscosity with increasing temperature would be expected to affect the rate of shrinkage most of all. Yet, a sharp increase of shrinkage with rise in temperature is noted in the final stage of shrinkage, which cannot be due to the viscosity of the metal. Thus, it is not the viscous flow of the liquid phase that controls the densification process.

Studies of the variation of the compressive force with the angle of contact have shown (see Fig. 60) that with decreasing angle of contact the capillary force increases sharply. For example, at $\varphi = 30°$, changing the angle of contact from $30°$ to zero increases the compressive force almost one and a half times. With rise in sintering temperature the wetting angle decreases sharply and as a consequence of this the capillary pressure − the driving force of the process − must increase sharply.

In the W−Cu system the angle of contact θ decreases to zero at a temperature of 1350°C and consequently the capillary pressure increases. Also, the dihedral angle [see Eq. (2)] necessary for the liquid to penetrate into the joints betw.. n the particles is attained. A large number of particles slip with respect to each other, and this also affects shrinkage.

Furthermore, as can be seen from Fig. 22, the greatest change in shrinkage occurs in a range of temperatures close to 1350°C, while above this temperature the shrinkage changes but little. This conforms with the variation of wetting. Indeed, at a temperature above 1350°C the angle of contact between tungsten and copper remains zero, i.e., the compressive force reaches its limiting value and does not increase further. Thus, for the W−Cu system, the shrinkage increases with rise in temperature as a consequence of a change in the angle of contact between tungsten and liquid copper.

Similar results were obtained for the tungsten−silver system. It is interesting that the curve of shrinkage increase with rise temperature is steeper for the W−Cu system than for the W−Ag system (in conformity with the sharper decrease of the angle of contact for the former system). Shrinkage in the W−Cu system is generally greater than in the W−Ag system (Fig. 73).

This can be attributed to the fact that higher compressive forces occur during the sintering of the W−Cu system due to the higher surface tension of copper by comparison with silver and to the smaller angle of contact. It was also found that the capillary pressure compressing the compact during sintering can vary greatly with the amount of liquid phase in the sample.

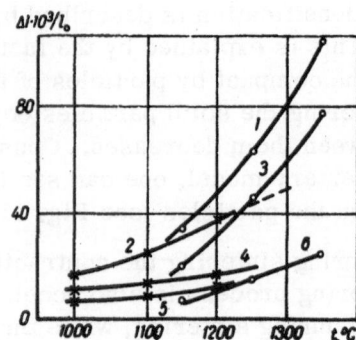

Fig. 73. Variation of shrinkage with sintering temperature in the W–Ag and W–Cu systems (initial porosity 40%). 1) W + 50 vol.% Cu; 2) W + 35.0 vol.% Cu; 3) W + 50 vol.% Ag; 4) W + 35 vol.% Ag; 5) W + 20 vol.% Ag; 6) W + 20 vol.% Cu.

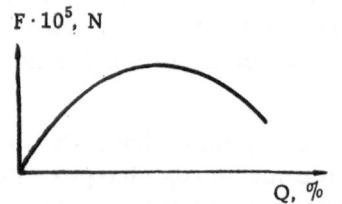

Fig. 74. Variation of compressive force with amount of liquid.

Let us examine the variation of the contracting force in a porous body with change in the degree of pore filling by the liquid. For particles of irregular shape, as used in our investigation (for example, the tungsten powder had a cubic shape), the capillary force should increase with increasing amount of liquid phase, beginning from zero (as shown in Chapter 7). However, as the pores are filled by the liquid the pressure-inducing meniscuses begin to disappear and the capillary force decreases, again reaching zero on complete filling of the pores. Thus, the variation of the compressive force with the amount of liquid phase should show a maximum (Fig. 74). It would appear that shrinkage as a function of the composition of the sintered substance will also exhibit an extreme point.

Measurement of shrinkage with impregnation and subsequent sintering as well as after conventional sintering in systems of various types such as W–Cu, W–Ag, and Cr_3C_2–Cu–Ni (see Figs. 23, 26, 41) shows that the shrinkage has a maximum at approximately 50% filling of the pores with the liquid phase. Evidently this amount of liquid induces the highest capillary force and is the most favorable for the displacement and regrouping of the particles.

The highest degree of densification during sintering is attained in the process of regrouping of the solid particles. The possibility of attaining complete densification with 35 vol.% of liquid phase in an alloy as a result of the regrouping process alone has not been confirmed, as indicated by Kingery, for systems with noninteracting components.

On the basis of our own work we conclude that complete densification as a result of the regrouping process alone can be obtained at liquid-phase concentrations reaching 50 vol.%, taking into account that in real systems the particles usually vary in size and are far from being spherical in shape.

The study of sintering processes in alloys with limited solubility of the components showed that the rate of the densification process in the initial stage of sintering, corresponding to the regrouping process, is described also by the relationship $\Delta l / l_0 \sim \tau^k$ (k = 1-4). The value of k depends on the sintering temperature and the composition of the alloy, unlike the theoretical value, which is close to unity.

In Kingery's theory it is assumed that the sintering temperature is reached instantaneously. Our data, as well as those from [65, 103], cannot be directly compared with Kingery's results [86] because of a difference in heating rates.

In our experiments the heating time from the temperature at which the liquid phase appeared to the given sintering temperature was 4-10 min. Therefore shrinkage could have begun during the actual heating, thereby introducing a certain inaccuracy in the measurement of sintering time.

From additional experiments undertaken to determine more precisely the slope of the curve in the regrouping process at very high heating rates (increasing the temperature from

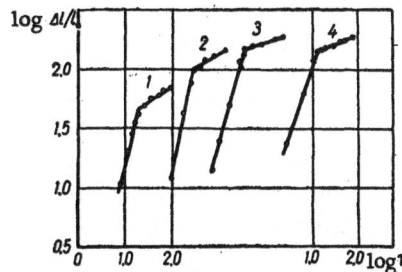

Fig. 75. Densification curves of the Ni–Tic system in logarithmic coordinates (sintering temperature 1350°C). 1) $k_1 = 1.7$, $k_2 = 0.27$ (TiC + 5 vol.% Ni); 2) $k_1 = 2$, $k_2 = 0.25$ (TiC + 17 vol.% Ni); 3) $k_1 = 1.6$, $k_2 = 0.2$ (TiC + 28 vol.% Ni); 4) $k_1 = 1.4$, $k_2 = 0.2$ (TiC + 35 vol.% Ni).

900 to 1400°C in 1.5–2 min) it can be seen that under such conditions k decreases to 1.4–2 (Fig. 75). The densification rate in the second stage of the process is small and is given by the relationship $\Delta l/l_0 \sim \tau^k$ (k = 0.1–0.5). The experimental results are in satisfactory agreement with the theoretical data given by Kingery for the second stage of sintering.

In the second stage of sintering the exponent k increases with decreasing amount of liquid phase in the sample. This is evidently due to the fact that with small amounts of liquid phase the contribution of the solution-and-precipitation process to the total densification increases, while with substantial amounts of liquid phase densification occurs to a large extent by the regrouping process and the solution-and-precipitation process has little effect on the total densification (shrinkage).

In the solution process the material is transported by way of the liquid phase and deposited in those areas where the grains are not in contact. The grains become adjusted to each other in shape, which promotes shrinkage. The development of a characteristic grain shape during liquid-phase sintering, particularly in cemented carbides, has attracted the attention of many investigators. The shape of the grains in sintered titanium carbide with nickel or cobalt is determined by the sintering conditions.

Whalen and Humenik [123] studied the sintering of titanium carbide with nickel and cobalt on backing plates of aluminum oxide, graphite, and titanium carbide in a vacuum, hydrogen, and helium and came to the conclusion that the equilibrium shape of the titanium carbide grains is a cube or rectangular prism with acute angles and sharp edges. In their opinion, rounded grains can be obtained only in the presence of oxygen.

The interfacial energy of titanium carbide grains at a boundary with liquid is different on different crystal faces [119]. Under the influence of oxygen the surface-energy anisotropy of the crystal faces decreases and grain growth becomes more isotropic, which promotes the formation of spherical grains.

In sintered samples of titanium carbide with nickel (see Fig. 34) the faceting of the grains also varies with change in the binder concentration. With large quantities of liquid (28 and 35 vol.% Ni) the grains are discrete and disseminated in a metallic matrix. At 17 vol.% Ni coalescence of some grains is observed. The grains are of angular shape. Thus, the shape of the grains varies depending on the amount of liquid phase in the sample during sintering. Similar results were obtained in [103]. This is due to the fact that with small amounts of liquid phase the solution and precipitation of carbide grains will be localized in small areas between the grains. Under these conditions saturation of the liquid will be rapidly attained, i.e., the grains will largely retain their original shape. With a sufficient amount of liquid the carbide grains can move freely in the liquid phase and become partially dissolved, and the dissolved substance is then deposited on the faces of large grains of low surface energy, which is responsible for the change in their shape.

Whalen and Humenik found no changes in the shape of carbide grains with the amount of binder in TiC–Ni and TiC–Co cermets, but this was evidently due to the fact that the experiments were made with only two compositions (22 and 33 vol.% binder), and consequently it was difficult to determine its effect, especially since the amount of binder was large in both cases.

In all cases one observes grain growth with increasing temperature. Raising the temperature to 1500°C promotes not only grain growth but also considerable grain-size equalization.

Metallographic analysis showed that in TiC–Ni and TiC–Co cermets no rigid skeleton is formed when the concentration of binder is 28 vol.% or higher. At lower metal concentrations bonds are formed between the carbide grains, their number depending on the amount of binder. With increasing amount of binder fewer bonds are formed, the grains may be bonded to other grains at only one or two points, and finally may become completely separate.

Densification during the sintering of components with limited solubility is substantially affected by an increase of temperature and the concentration of liquid phase in samples [36, 59]. Thus, in samples of the Co–TiC and Ni–TiC systems with binder concentrations of about 30 vol.% the residual porosity does not exceed 10% at a temperature of 1350°C or higher; at low binder concentrations (up to 20 vol.%) pore-free samples are not obtained even at 1500°C.

With rise in temperature, when the amount of liquid metal increases due to the increase in the solubility of the high-melting component, the contribution of the regrouping process to the total densification increases. For example, at a temperature of 1500°C, this contribution amounts to about 80 vol.% of the total shrinkage in Ni–TiC alloys, although complete densification is attained at this temperature only with an initial concentration of the low-melting component above 30%; according to the phase diagram at 1500°C this corresponds to the formation of around 50 vol.% of liquid. At lower temperatures (up to 1400°C) pore-free samples are not obtained on sintering for up to 2 h.

An increase of the sintering time promotes shrinkage, but the alloy microstructure is then affected by the solution-and-precipitation process. Therefore, to obtain a fine-grained structure and strong sintered material a high sintering temperature and a short sintering time are desirable.

Investigation of the effect of porosity on densification in systems with mutually soluble components showed that the higher the initial porosity of samples the greater is their shrinkage. However, the effect of porosity on total densification is small, in contrast to systems with mutually insoluble components, where the density of sintered samples increases with increasing initial porosity.

On the basis of investigations of the Cr_3C_2–Cu–Ni system it can be concluded that, as in other carbide systems, the densification process in this system also occurs in two stages governed by the relationship $\Delta l / l_0 \sim A \tau^k$ and the exponents are similar.

Judging from literature data on the reaction of chromium carbide with copper and on the angle of contact, the densification process in the Cr_3C_2–Cu system must result from the regrouping process. In this system the shrinkage process is similar to that in the W–Cu and W–Ag systems. The increase of shrinkage with rise in temperature for Cr_3C_2–Cu alloys is also due to the effect of temperature on the angle of contact. When nickel is added to the copper the wetting of the chromium carbide improves and its solubility in the copper–nickel alloy increases. It can be seen from Fig. 38 that shrinkage increases with increase in temperature and the amount of nickel added to the Cu–Ni alloy. In alloys containing from 6 to 10 vol.% Ni, however, the angle of contact decreases to zero with increasing temperature. In this case one also observes a sharp change in sample density on the densification curves (Fig. 40). This underlines the important role of wetting not only in systems in which the components are mutually insoluble but also in systems with limited solubility of components.

The results of investigations into the sintering of samples from mixtures of nickel and lead show that densification in this system is small. A rise in sintering temperature also promotes shrinkage by increasing the solubility of nickel in lead. However, the shrinkage in this system is small in comparison with that in similar systems. A substantial increase in the amount of liquid phase also has little effect on the shrinkage.

The angle of contact of nickel with lead is close to 50° and changes only little in the temperature range of 400-800°C [14]. Evidently, as a result of the low degree of wetting, the energy relations at the particle junctions are such that the liquid phase cannot penetrate between the nickel grains and collects in separate drops in the sample pores, mainly occupying the positions designated as "b" and "c" in Chapter 2. Hence, an increase in the relative amount of liquid not only does not result in increased shrinkage but leads to sweating out of the liquid phase. Thus, at a temperature of 850°C and 28 vol.% Pb it was found impossible to measure shrinkage owing to sweating out of the liquid phase and swelling of samples.

It is impossible to obtain dense nickel—lead alloy samples by sintering with liquid-phase participation. With rise in sintering temperature marked grain growth is observed in the nickel (the grain diameter increases 10- to 15-fold).

In view of the low solubility of the components of this system, such large growth of the nickel grains cannot be ascribed to the solution-and-precipitation process. It can be assumed that in this case, too, at a large angle of contact the dihedral angle is substantial, which does not allow the liquid to penetrate between the grains and ensures "dry" contact between the nickel grains. Therefore, the nickel grains merge together (coalesce) during heating, the possibility of which was indicated by Humenik and Parikh [108]. According to Humenik and Parikh, coalescence begins when the angle of contact is substantially larger than zero but less than 90°. The coalescence and growth of the solid-phase grains at the points of contact are promoted by the solution process with subsequent precipitation. Makarova et al. [30] investigated the effect of grain misorientation on the coalescence process and showed that, in the absence of misorientation of adjacent crystals, i.e., when $\sigma_{S_1-S_2} = 0$, coalescence occurs. When grains with different spatial orientations are drawn together, the interphase boundary is thermodynamically more stable.

In recent years much work has been done on the activation of sintering, which can be attained by different methods: by oxidation—reduction reactions, by applying a magnetic field or ultrasonic vibrations, etc. The most effective in this respect are various additions promoting the formation of a liquid phase.

It can be seen from work done on model systems that in the case of spherical particles even a very small amount of liquid will generate substantial capillary forces; with increasing amount of liquid these forces at first decrease sharply and then change very little. Hence, very small additions are required for the activation of the process.

If the particles are of irregular shape, the capillary force increases with increasing amount of liquid. Therefore, for powders with particles of irregular shape a large amount of liquid is preferable. In this and other cases improved wetting of the solid phase by the liquid leads to larger capillary forces. For the activation of the sintering process it is also desirable that the additions have a high surface tension.

LITERATURE CITED

1. L. B. Al'tman and V. L. Memelov, Poroshkovaya Metallurgiya, No. 6, pp. 44-54 (1961).
2. Yu. N. Artyukh, V. N. Eremenko, and V. P. Solomko, Information Bulletin 116 [in Russian], IMSS AN UkrSSR (1958).
3. N. K. Adam, Physics and Chemistry of Surfaces [Russian translation], OGIZ, Gostekhizdat (1947).
4. M. Yu. Bal'shin, Powder Metallurgy [in Russian], Mashgiz, Moscow (1948).
5. A. S. Berezhnoi, Ogneupory, 8:351 (1948).
6. Z. V. Volkova, Zh. Tekh. Fiz., 13:224 (1939).
7. Ya. E. Geguzin, Dokl. Akad. Nauk SSSR, 124(5):1045 (1959).
8. V. N. Goncharova, Metalloved. i Term. Obrabotka Metal., No. 3, pp. 51-53 (1955).
9. V. V. Grigor'eva and V. N. Klimenko, Chromium Carbide Alloys [in Ukrainian], Vid. AN UkrRSR, Kiev (1961).
10. B. V. Deryagin and A. A. Zimon, Kolloidn. Zh., 23(5):544 (1961).
11. B. V. Deryagin and N. A. Krotova, Adhesion [in Russian], Izd. AN SSSR (1949).
12. V. N. Eremenko and I. A. Lavrinenko, Information Bulletin 218 [in Russian], IMSS AN UkrSSR (1960).
13. V. N. Eremenko and N. D. Lesnik, Vopr. Poroshkovoi Met. i Prochnosti Materialov, No. 3, p. 73 (1956).
14. V. N. Eremenko and N. D. Lesnik, in: Surface Phenomena in Metals and Alloys and Their Role in Powder Metallurgy Processes [in Russian], Izd. AN UkrSSR, Kiev (1961), pp. 155-177.
15. V. N. Eremenko and Yu. V. Naidich, Surface Wetting of Refractory Alloys by Liquid Metals [in Ukrainian], Vid. AN UkrRSR, Kiev (1958).
16. V. N. Eremenko and Yu. V. Naidich, Zhur. Neorg. Khim., 4(9):2052 (1954).
17. V. N. Eremenko, Yu. V. Naidich, and I. A. Lavrinenko, Poroshkovaya Metallurgiya, No. 4, p. 72 (1962).
18. V. N. Eremenko, Yu. V. Naidich, and I. A. Lavrinenko, in: Proceedings of the Seventh Scientific-Technical Conference on Powder Metallurgy, [in Russian], Erevan (1964).
19. V. N. Eremenko and Ya. V. Natanzon, Metalloved. i Term. Obrabotka Metal., No. 1, pp. 39-42 (1960).
20. V. N. Eremenko and Z. I. Tolmacheva, Poroshkovaya Metallurgiya, No. 2, p. 21 (1961).
21. O. A. Esin, P. V. Gel'd, and S. I. Popel', Dokl. Akad. Nauk SSSR, 74(6):1087 (1950).
22. N. M. Zarubin and R. A. Trubnikov, Redkie Metally, No. 2, p. 38 (1935).
23. A. D. Zimon, Kolloidn. Zh., 25(3):317 (1963).
24. T. N. Znatokova and V. I. Likhtman, Fiz. Metal. i Metalloved., 4(3):511-518 (1957).
25. V. L. Itin et al., Izv. Vysshikh Uchebn. Zavedenii, Fizika, No. 2, p. 139 (1965).
26. I. P. Kislyakov, in: Surface Phenomena in Melts and Powder Metallurgy Processes [in Russian], Izd. AN UkrSSR, Kiev (1963), p. 182.
27. P. Schwartzkopf and R. Kieffer, Refractory Hard Metals, Macmillan, New York (1953).

28. V. I. Likhtman and L. T. Nazarov, Zh. Tekhn. Fiz., 22(4):696 (1952).
29. A. V. Lykov, Transport Phenomena in Capillary-Porous Bodies [in Russian], Gostekhizdat, Moscow (1954).
30. R. V. Makarova, O. K. Teodorovich, and I. N. Frantsevich, Poroshkovaya Metallurgiya, No. 7, p. 45 (1965).
31. R. S. Mints, Dokl. Akad. Nauk SSSR, 118(3):543-545 (1958).
32. Yu. V. Naidich and V. N. Eremenko, Fiz. Metal. i Metalloved., 1(6):883 (1961).
33. Yu. V. Naidich and G. A. Kolesnichenko, Poroshkovaya Metallurgiya, No. 6, p. 55 (1963).
34. Yu. V. Naidich and G. A. Kolesnichenko, Poroshkovaya Metallurgiya, No. 1, p. 61 (1963).
35. Yu. V. Naidich and G. A. Kolesnichenko, Poroshkovaya Metallurgiya, No. 3, p. 23 (1964).
36. Yu. V. Naidich, I. A. Lavrinenko, and V. N. Eremenko, Poroshkovaya Metallurgiya, No. 1, p. 5 (1964).
37. Yu. V. Naidich, I. A. Lavrinenko, and B. Ya. Petrishchev, Poroshkovaya Metallurgiya, No. 2, p. 50 (1965).
38. Yu. V. Naidich and I. A. Lavrinenko, Poroshkovaya Metallurgiya, No. 10, p. 61 (1965).
39. Yu. V. Naidich and I. A. Lavrinenko, in: Surface Phenomena in Melts and the Solid Phases in Them [in Russian], Kabardino-Balkarskoe Knizhnoe Izd. Nal'chik (1965).
40. B. Ya. Pines, Zh. Tekhn. Fiz., Vol. 18, No. 8 (1957).
41. B. Ya. Pines, Usp. Fiz. Nauk, 52:501 (1954).
42. G. I. Pokrovskii, Capillary Forces in Soils [in Russian], Gosstroiizdat, Moscow (1933).
43. K. V. Savitskii et al., Poroshkovaya Metallurgiya, No. 11, p. 19 (1965).
44. V.K. Semenchenko, Surface Phenomena in Metals and Alloys [in Russian], Gostekhizdat, Moscow (1957).
45. V. L. Tret'yakov, Sintered Hard Alloys [in Russian], Gostekhizdat, Moscow (1962).
46. N. A. Fuks, Progress in the Mechanics of Aerosols [in Russian], Izd. AN SSSR, Moscow (1961).
47. G. I. Fuks, in: Collection of Reports on the Industrial Efficiency of Petroleum, No. 1 [in Russian], Gostekhizdat (1948), p. 266.
48. M. Hansen and K. Anderko, Constitution of Binary Alloys, McGraw-Hill, New York (1958).
49. A. A. Shmykov, Poroshkovaya Metallurgiya, No. 4, p. 32 (1956).
50. N. J. Ananthanarayanan and J. E. Gibsch, J. Metals, No. 5, pp. 78-80 (1953).
51. R. K. Beggs, J. Metals, 3(1):860-865 (1951).
52. R. Bernard, Plensee Proceedings 1955, Pergamon Press, London (1956), pp. 41-49.
53. A. Bondi, Chem. Rev., 52(2):417-458 (1953).
54. B. Bovarnick, Proc. Metal Powder Indust. Federation (1959), pp. 72-81.
55. R. Bradley, Phil. Mag., 6:695 (1958); 13:853 (1932).
56. C. Broquet, R. Margerand, and M. Eudier, Mem. Sci. Rev. Met., 60(3):171-176 (1963).
57. C. Broquet, Rev. Nickel, 29(2):36-45 (1963).
58. N. Cabrera, Trans. AIME, 188:668 (1950).
59. H. S. Cannon and F. V. Lenel, Pulvermetallurgie, Plansee Seminar De re metallica, Springer-Verlag, Vienna (1953), p. 106.
60. B. Cech, Hutnicke Listy, 11(7):419-424 (1956).
61. B. Cech, Konferencja Metallurgii Proszkow (1963), Polska Akademia Nauk, Cracow.
62. B. Cech, J. Powder Met., 1:112 (1963).
63. B. Cech, Hutnicke Listy, 15(4):287 (1960).
64. B. Cech, Neue Hütte, 3(5):300 (1953).
65. B. Cech, VUPM, 4:1 (1964).
66. V. W. Cegielski, Proceedings of the Second International Powder Metallurgy Congress, Eisenach (1961), Vols. 1-3, pp. 109-119.
67. W. Dawihl and J. Hinnuber, Z. Kolloid., 104:233 (1943).
68. P. Duwez and H. Martens, J. Metals, 1:571-577 (1949).

69.   C. N. Davies, Interaction of Water and Porous Materials, No. 3, p. 123 (1948).
70.   R. Edwards and T. Raine, Pulvermetallurgie, Plansee Seminar de re metallica, Springer-Verlag, Vienna (1953), p. 235.
71.   J. E. Elliot, Metallurgia, 59(351):27–37 (1959).
72.   J. McFarlane and D. Tabor, Proc. Roy. Soc., 202(1069):224–243 (1950).
73.   K. Forkar and U. Cotz, Z. Metallk., 46(5):371–377 (1955).
74.   J. Gurland and J. T. Norton, J. Metals, 4(10):1040 (1952).
75.   J. Gurland and J. T. Norton, Second Plansee Seminar De re metallica, Reutte Tyrol., Metallwerk Plansee, pp. 99–100 (1955).
76.   J. Gurland, Trans. AIME, 215:601–608 (1959).
77.   J. Gurland, Jernkont. Ann., 147:1 (1963).
78.   J. Gurland, J. Metals, 6(2):285 (1954).
79.   J. Graham, J. H. Weymouth, and L. S. Williams, J. Austral. Inst. Met., 8(3):280–297 (1963).
80.   J. Gurland, Trans. AIME, 215(4):601–608 (1959).
81.   H. H. Hausner, Konferencja Metalurgii Proszkow (1963), Polska Akademia Nauk, Cracow.
82.   M. Humenik and N. M. Parikh, J. Am. Ceram. Soc., 39:60 (1956).
83.   J. Hinnuber and O. Rudiger, Archiv Eisenhuttenw., 5–6:267–274 (1953).
84.   C. Kawashima, S. Saito, and T. Hanazawa, J. Ceram. Assoc. Japan, 66(9):191 (1958).
85.   W. D. Kingery, J. Appl. Phys., 3:301 (1959).
86.   W. D. Kingery and M. D. Narasimhan, J. Appl. Phys., 3:306 (1959).
87.   W. D. Kingery and M. D. Narasimhan, J. Am. Ceram. Soc., 44(1):29–35 (1961).
88.   W. D. Kingery and M. Berg, J. Appl. Phys., 26:1205 (1955).
89.   G. C. Kuczynski, J. Metals, 1(Sec. II):169 (1949).
90.   A. Krupkowski, W. Rutkowski, and S. Stolarz, Glown. Inst. Met., Prace, No. 3, pp. 297–305 (1951).
91.   W. Koster and J. Raffelsieper, Z. Metallk., 42:387–391 (1951).
92.   G. Kofler, Proceedings of the First International Powder Metallurgy Congress, Groz (1948), Vol. 49, p. 187.
93.   N. S. Kothari and J. Waring, Powder Metallurgy, 7:13 (1964).
94.   F. V. Lenel, Trans. AIME, 175:878–896 (1948).
95.   F. V. Lenel, The Physics of Powder Metallurgy, McGraw-Hill, New York (1951), pp. 238–255.
96.   O. Langraf, Metall., 3(11–12):184–186 (1949).
97.   E. H. Lewin, F. McMurdie, and F. P. Hall, Phase Diagrams for Ceramists, Am. Ceram. Soc., Ohio (1956).
98.   S. Makipirtti, Metal Powder Report, 2:32 (1960).
99.   J. T. Norton, Powder Metall. Bull., 6:75 (1951).
100.   J. T. Norton, J. Metals, 8(1):49–53 (1956).
101.   E. Niki, S. Konara, and M. Tajiri, J. Ceram. Assoc. Japan, 69(6):169 (1961).
102.   E. Niki, S. Konara, and K. Tatsuzawa, J. Ceram. Assoc. Japan, 70(11):313–318 (1962).
103.   E. Niki et al., Trans. Japan Inst. Metals, 5(1):1 (1964).
104.   J. Okamoto, J. Japan Soc. Powder Metallurgy, 9(1):1–6 (1962).
105.   T. Price, C. Smithells, and S. Williams, J. Inst. Met., 62(1):239–269 (1938).
106.   W. Plate, Z. Metallk., 42(3):761 (1951).
107.   D. G. Protopopescu, Rev. Roumaine Met., 5(1):107–117 (1960).
108.   N. M. Parikh and M. Humenik, J. Am. Ceram. Soc., 40(9):315 (1957).
109.   E. Pelzel, Metall, 9:692–694 (1955).
110.   H. R. Peiffer, Trans. AIME, 218(5):755 (1960).
111.   E. Raub and W. Plate, Z. Metallk., 40(6):206–214 (1949).
112.   E. Raub and W. Plate, Z. Metallk., 40(5):171–175 (1949).

113. C. S. Smith, Trans. AIME, 175:17 (1948).

114. W. P. Sykes, The Physics of Powder Metallurgy (1951), p. 14.

115. L. Skolnick, Kinetics of High Temperature Processes, New York-London (1959), p. 92.

116. K. Schreiner and G. Glawitsh, Z. Metallk., 45(3):102-108 (1954).

117. E. G. Sandford and E. M. Trent, The Iron and Steel Institute, Special Report, 38:84 (1947).

118. W. Stone, Phil. Mag., 9(58):610 (1930).

119. L. Skolnick, J. Metals, 9(4) (Sec. II):438-442 (1957).

120. R. B. Shaw et al., Trans. Am. Soc. Met., 45:249 (1953).

121. W. J. O'Brien and C. J. Ryge, J. Am. Ceram. Soc., 47(1):5 (1964).

122. J. Weyl, Ceramic Age, 60(5):28 (1952).

123. J. Whalen and M. Humenik, Trans. AIME, 218:5 (1960).

124. T. Gillespie and W. Settinery, J. Coll. Interf. Sci., 24:199 (1967).

125. Discussion, J. Coll. Interf. Sci., 25:246 (1968).